# The Physician's
# Golf Injury
## Desk Reference

JONES AND BARTLETT PUBLISHERS
*Sudbury, Massachusetts*
BOSTON   TORONTO   LONDON   SINGAPORE

## Dr. Jeff Blanchard

*World Headquarters*
Jones and Bartlett Publishers
40 Tall Pine Drive
Sudbury, MA 01776
978-443-5000
info@jbpub.com
www.jbpub.com

Jones and Bartlett Publishers
Canada
6339 Ormindale Way
Mississauga, Ontario
L5V 1J2
CANADA

Jones and Bartlett Publishers
International
Barb House, Barb Mews
London W6 7PA
UK

Jones and Bartlett's books and products are available through most bookstores and online booksellers. To contact Jones and Bartlett Publishers directly, call 800-832-0034, fax 978-443-8000, or visit our website www.jbpub.com.

Substantial discounts on bulk quantities of Jones and Bartlett's publications are available to corporations, professional associations, and other qualified organizations. For details and specific discount information, contact the special sales department at Jones and Bartlett via the above contact information or send an email to specialsales@jbpub.com.

6048

**Library of Congress Cataloging-in-Publication Data**

Blanchard, Jeff.
The physician's golf injury desk reference / Jeff Blanchard.
p. ; cm.
ISBN-13: 978-0-7637-4613-1
ISBN-10: 0-7637-4613-4
1. Golfers—Health and hygiene—Handbooks, manuals, etc. 2. Golf injuries—Prevention—Handbooks, manuals, etc. 3. Sports injuries—Handbooks, manuals, etc. I. Title.
[DNLM: 1. Athletic Injuries—diagnosis. 2. Golf—physiology. 3. Athletic Injuries—etiology. 4. Athletic Injuries—therapy. 5. Biomechanics.
QT 260.5.G6 B639p 2007]
RC1220.G64B553 2007
617.1'027024796352—dc22
2006022694

**Production Credits**
Executive Editor: David Cella
Production Director: Amy Rose
Associate Production Editor: Rachel Rossi
Editorial Assistant: Lisa Gordon
Manufacturing Buyer: Amy Bacus
Composition: Pre-Press Company, Inc.
Cover Design: Kristin E. Ohlin
Printing and Binding: Malloy, Inc.
Cover Printing: John P. Pow Company

Printed in the United States of America
10 09 08 07 06    10 9 8 7 6 5 4 3 2 1

# Dedication

A special thank you to Chantal Blanchard and Mark Deutman for their help, for their expertise, and for their patience with me throughout the publishing of this book.

# Contents

# CHAPTER 2

## SWING MECHANICS
### Three Key Points

# CHAPTER 3

## BARRIERS TO PERFORMANCE
### Well Patient Evaluations and Management Protocols

# PART 2

# FOR THE PLAYER

# CHAPTER 4

## TRAINING FOR GOLF

### Strength, Power, and the 9-Hole Playing Lesson

# CHAPTER 5

# FEAR AND FREEDOM
## The Golfer's Guide to Mental Health

# Preface

Golf is an expanding market. Currently, about 27.5 million men, women, and children play golf in the United States. There are more than 40 million golfers worldwide. In 1950, there were about 5,000 golf courses. Today there are more than 15,000 golf courses, and some 400 to 500 more are being added yearly.

> "Golf has grown faster than motion pictures, financial services, hotels and communications… all fast growing industries."
> — *National Golf Foundation*

As a physician, are you positioned in the marketplace? Do you understand golf has developed into an athletic event in which nearly 50 percent of golfers eventually suffer from a golf-related injury? Do you realize that the typical golf patient is highly motivated to improve his or her game? Golfers are, in general, influential and affluent networkers who can afford treatment and who usually refer their friends. These are an educated and appreciative people who recognize an expert when they meet one. Are you the expert they are looking for?

You don't even need to like golf. What you need to know are the three root causes of golf injuries, how to give a comprehensive physical examination that is golf-specific, and the treatment and conditioning protocols for your injured golf patient.

> You do not have to play golf to treat golfers.

The average golf patient does not prepare or condition his or her body for golf. These people just play. They will show up minutes before their tee time and bash it about for 18 holes of golf. If they can't play a round of golf, they will go to a driving range and hit a couple hundred golf balls. No warm-up, no stretching—just playing golf or just hitting balls. Ball after ball after ball.

The average patient can swing the golf club 90 miles per hour (mph). The average tour pro swings the club about 115 mph. Paul Chek, in his book *The Biomechanics of Golf* (1999), writes: "Amateur golfers achieve approximately 90 percent of their peak muscular activity when driving a golf ball. This is the same intensity as picking up a weight that can only be lifted four times before total fatigue."

And that's just one swing! Even though par on a regulation 18-hole golf course is 72 strokes, the average golfer needs more than 100 strokes. Assuming half of your patients' strokes are of lesser intensity (chipping and putting), this means your patients are swinging with "all their might" at the ball about 50 times over a 5-hour round of golf. To make matters worse, there is often a lot of downtime between swings—time spent looking for wayward golf shots, time spent waiting for the slow groups ahead to move out of the way—and most of your patients are using golf carts instead of walking. It's easy to understand why a patient's body cools off and stiffens up between swings. The average patient has no warm-up or stretching protocols for golf. Your patients are swinging the club with violent, intermittent effort. If ever there was a recipe for injury, this is it.

In addition to intermittent and the "grip it and rip it" golf swings, many of your patients have varied amounts of preexisting postural dysfunction and poor flexibility. When you add it all up, it's no wonder golfers need your help.

The golf injury is but another variety of a repetitive strain syndrome. You already know how to treat sprains and strains. What might be new to you is the diagnosis of the golf swing. Talk about walking onto thin ice! Your patients are suffering from repetitive strain injuries caused by poor swing mechanics.

Your patients are swinging the club in such a way that they are putting undo and often extreme strain on muscles, tendons, ligaments, joints, and bones. They swing often and they swing with a determined violence to control the flight path of the ball through the air.

In the following pages, I have made the golf swing very simple to diagnose. There are three key points to master. Once understood, you will be able quickly and efficiently to diagnose and treat the root cause of the repetitive strain syndrome. I introduce a new language of golf terminology that is based on the simple principles of geometry and the biomechanics of human movement. This is a language with which we are generally familiar and which we can apply to golf. We are not competing with golf professionals giving golf lessons to patients. We are health care providers treating injured golfers. It is our responsibility to identify the root cause of the repetitive strain syndrome. When a patient's injuries are caused by poor swing mechanics, we use neuromuscular reeducation to help our patients swing the club safely and efficiently.

Have fun with this. If you play golf, become your own first patient. Apply the principles, training, and conditioning models presented here to your own game. I have no doubt you will soon be playing better golf as well as helping your patients not only heal their present golf-related injuries, but prevent new ones as well as improve their game.

# Introduction

You don't have to play golf to treat golfers. You don't even need to like golf. What you need to know are the three root causes of golf injuries, how to give a comprehensive physical examination that is golf-specific, and the treatment and conditioning protocols for your injured golf patients.

There are three fundamental causes of golf injuries: postural instability, lack of flexibility, and poor swing mechanics. The root cause of poor swing mechanics is often the result of physical restrictions and related mechanical adhesions that interfere with the normal movement of bone and joint. When combined with a general lack of golf-specific flexibility and golf-specific postural stability, it's no wonder your patients are at risk of injury.

On the golf course, physical exertion is intermittent. A golfer will attempt approximately 50 to 70 violent swings every 5 minutes or so while playing 18 holes. The average amateur swings the club at 80 to 100 miles per hour. On the driving range, the pounding is rapid and relentless. Your patient will flail away at golf balls 60 to 100 times in a half hour, and often continue the assault for hours on end. Patients who become injured playing golf require a golf-specific physical examination.

In *The Golf Biomechanics Manual*, Paul Chek (1999) writes: "Amateur golfers achieve approximately 90 percent of their peak muscular activity when driving a golf ball. This is the same intensity as picking up a weight that can only be lifted four times before total fatigue. This level of exertion and muscular activation equates golf with such sports as football, hockey and martial arts. The difference is that other athletes outside of golf include conditioning as an integral part of their preparation before play." Very few golfers attempt any conditioning at all. Is it any wonder there are so many golf-related injuries? The golf swing requires the spine to rotate, bend laterally, and extend. This requires flexibility that is golf-specific.

It is not your job to teach the golf swing but to identify the biomechanical movement responsible for a repetitive strain injury. It is up to you to help your patients "reroute" their swing path to prevent further injury.

Remember, you do not have to play golf to treat injured golfers.

In this text, I present a golf-specific examination procedure that can help you identify potential problem areas for your patients. Once a problem is identified, you will learn the appropriate correction sequences. You will learn appropriate clinical examination techniques and essential treatment protocols.

This book presents neuromuscular reeducation, golf fitness, and golf conditioning, not a program of golf instruction. You are not competing with the local PGA teaching professional, but you will learn to work along with the golf pro to the benefit of your patients. This methodology is leading-edge management for golf-related pain and injury. You will be able to position yourself as an expert in the diagnosis and treatment of golf injuries. The result is that the PGA teaching professionals will seek your help with their students who have pain stemming from postural dysfunction and poor flexibility.

# Part I

## FOR THE PHYSICIAN

# Chapter 1

## PHYSICAL EXAMINATION

### Structural and Functional Evaluation

The Anatomy of a Golf Injury

Physical Examination

14-Point Flexibility Examination

Golf Posture

Patient Swing Analysis

## THE ANATOMY OF A GOLF INJURY

The typical amateur golfer can swing a golf club 90 miles an hour. When your patients swing their clubs out of position or out of balance, they put tremendous strain on their joints, muscles, ligaments, and tendons. When the force applied to the tissues exceeds the tissues' resistance, they will tear (Figure 1-1).

We call tissue tears *strains* or *sprains*, depending on their severity. Soft-tissue injuries can heal in 6 to 8 weeks, depending on the severity of the injury and the age of the patient. However, when soft tissue heals, it becomes weaker than it was before the injury occurred. The injury site will mend with scars you cannot see. Scar tissue (fibrocytes) will infiltrate and replace the original tissue in and around the injury site.

> Fibrocytes are weaker, less elastic, and more pain-sensitive than original soft tissue is.

When challenged by the movement of the body, fibrocytes will not lengthen and give way; instead, they get angry, become irritated and inflamed, and restrict normal joint motion.

### CLINICAL MANAGEMENT: ACUTE VS. CHRONIC INJURY

Golf injuries fall into two general categories: acute and chronic. Acute injuries are new injuries and are usually associated with painful swelling. An acute injury prevents you from playing golf altogether. The first 4 to 6 weeks are crucial in the management/treatment of a new injury. In addition to pain relief, the goal of treatment is to reorganize the developing fibrocytes so they are more compatible with the surrounding tissue.

No treatment can prevent the formation of fibrocytes; however, left unattended, fibrocytes will form into an ugly, tangled mess that is considered a nonfunctional scar. To a patient, this means increased sensitivity to pain and greater loss of normal joint motion.

Golfers with acute injuries should get complete rest. Ice packs are recommended to reduce painful swelling. The ice can be applied for as long as 20 minutes per hour—in other words, 20 minutes "on," then 40 minutes "off." Repeat as often as necessary during the first 24 to 48 hours of an acute injury; follow up with a treatment program specifically designed to reorganize the fibrocytes into a "functional" scar. As a treating physician, you have

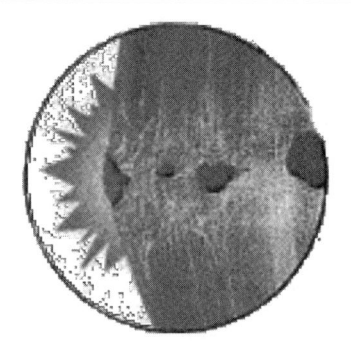

**FIG 1-1** Tear of the Myofascial Sheath

*Source*: Reprinted courtesy of the American Chiropractic Association.

several excellent soft-tissue therapy techniques from which to choose. If you do not know them, refer your patients to a provider who does. The long-term benefit will be less pain and a greater range of joint motion for the patient.

Golfers with chronic injuries, on the other hand, have a unique problem to overcome. Old injuries are often infiltrated with dense clusters of nonfunctional fibrocytes: trigger points (Figure 1-2). With overuse, trigger points inflame and feel like a new injury again. Usually, chronic injuries are associated with stiffness. A golfer with chronic injuries changes his or her swing to avoid pain. It's hard to play consistent, powerful golf with chronic injuries.

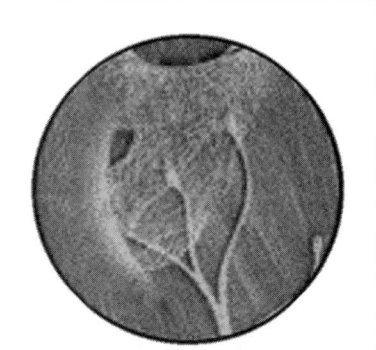

**FIG 1-2** Trigger Point

*Source*: Reprinted courtesy of the American Chiropractic Association.

## CLINICAL MANAGEMENT: STRAIN VS. SPRAIN

**STRAIN** When your patients injury muscle tissue, they are said to have strained the muscle. There are three general classifications of sprains:

### Minor
Muscle pain, muscle spasm, and/or muscle weakness.

### Moderate
Muscle pain and a general weakness of the muscle with movement.

### Severe
Partial or complete tear of the muscle tissue and/or tendon; movement is disabling.

**TREATMENT PROTOCOLS** The general rules of thumb for treatment of muscular strains include the following:

**Rest:** Reduce regular exercise or activities of daily living as needed. If movement is necessary, support is recommended, that is, tape, brace, sling, and/or crutches.

**Ice:** Apply an ice pack to the injured area for 20 minutes at a time. A cold pack, ice bag, or plastic bag filled with crushed ice and wrapped in a towel can be used. To avoid cold injury and frostbite, do not apply the ice for more than 20 minutes (Figure 1-3).

**Compression:** Compression of an injured ankle, knee, or wrist may help reduce swelling. Examples of compression bandages are elastic wraps, special boots, air casts, and splints.

**Elevation:** If possible, keep the injured area above the level of the heart to help decrease swelling.

**SPRAIN** When your patients injure the ligaments that tie bone to bone, they are said to have incurred a sprain. It is not uncommon for patients to hear a "pop" or a feel a "tear" at the moment of injury. As with muscle strains, there are classifications specific to the severity of the injury:

### Grade I

Slight tear with no bruise or joint instability. A person with a mild sprain usually experiences minimal pain, swelling, and little or no loss of functional ability. The person is usually able to put weight on the affected joint.

### Grade II

Partial tear characterized by bruising, moderate pain, and swelling. A person with a moderate sprain usually has some difficulty putting weight on the affected joint and experiences some loss of function.

### Grade III

Severe sprain—complete tear or rupture of a ligament. Pain, swelling, and bruising are usually severe, and the patient is unable to put weight on the joint. An x-ray is usually taken to rule out fracture.

**TREATMENT PROTOCOLS** The following protocols (described earlier) apply for treatment of sprains.

**R**est

**I**ce

**C**ompression

**E**levation

**Note:**
Gradual swelling indicates a chronic or long-lasting condition, whereas sudden swelling can signal a traumatic injury or an infection.

The amount of rehabilitation and the time needed for full recovery after a sprain or strain depend on the severity of the injury and individual rates of healing. For example, a moderate ankle sprain might require 3 to 6 weeks of rehabilitation before a person can return to full activity. With a severe sprain, it can take 8 to 12 months before the ligament is fully healed.

**BURSITIS** The bursa is a fluid-filled sac. Bursas are necessary in the body for the skin to be able to slide over the bone in a given area. It could be stated that the bursa acts as a cushion between the skin and the bone.

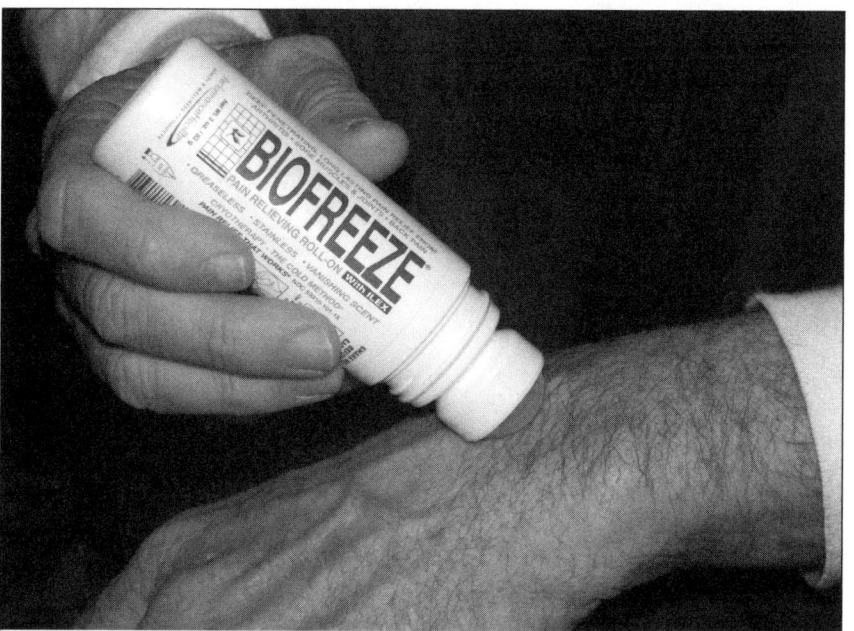

**FIG 1-3** In the absence of ice, topical application of the South American herb ILEX, found in formulations like BioFreeze, will reduce swelling and provide relief of pain.

Bursitis is an inflamed bursa. When injured, the bursa can collect more fluid, which is experienced as inflammation.

## CASE STUDY: ELBOW PAIN

The most common cause of bursitis of the elbow for those who play golf is a collapse of the forward arm (bending the elbow) during the backswing (Figures 1-5 and 1-6). During the downswing, the bent elbow snaps into extension as the club head impacts the ball. During 18 holes of golf, this could happen hundreds of times.

### TREATMENT PROTOCOL

**ACUTE:** Apply the **RICE** protocol to the inflamed bursa:

**R**est

**I**ce

**C**ompression

**E**levation

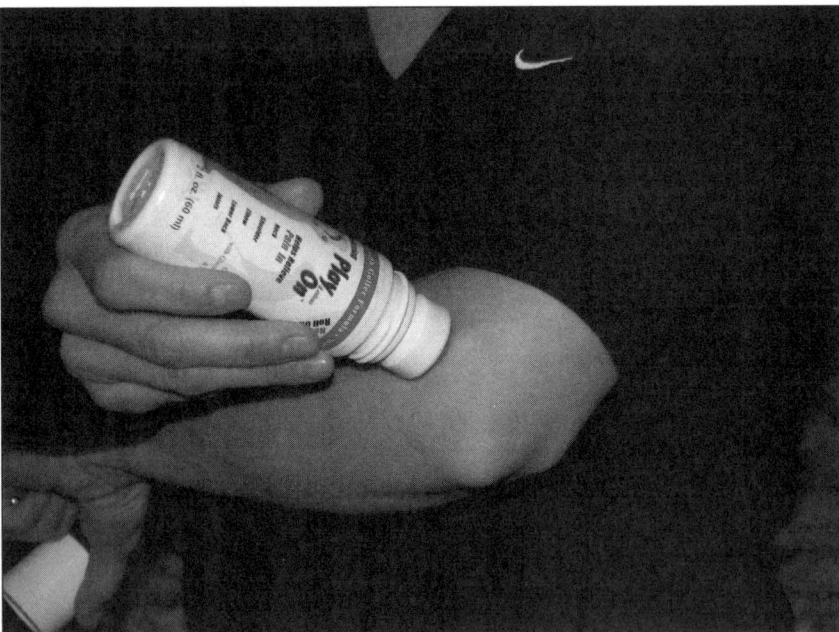

**FIG 1-4** Connective tissue associated with chronic strains and sprains needs to be "softened" before activity. In addition to stretching, have your patients apply a topical analgesic formula with the Indian spice tumeric. Play On Pain Relief Lotion has a formula that will stay warm for 4–6 hours per application.

**CHRONIC:** When practicing or playing golf, place tape across the forward elbow joint. The moment the elbow begins to collapse, the adhesive will pull the skin as a reminder to keep the forward elbow extended when swinging the club.

**TENDONITIS** Tendonitis of the wrist (also known as stenosing tenosynovitis) is an irritation and swelling of the sheath or tunnel that surrounds the tendons of the thumb.

Pain may increase with pinching or grasping with the fingers. Pain generally increases with flexion/extension of the hand relative to the forearm.

**WRIST** Tendonitis of the wrist is most likely to occur when striking golf balls that are sitting in deep rough or deep sand. Upon impact, there is rapid deceleration of the club head relative to the grip of the club, which "jams" the wrist.

> **Finkelstein Test**
>
> *Have your patient make the hand into a fist with the thumb tucked in toward the little finger. It is an abnormal finding if the test makes the pain worse.*

**FIG 1-5** To swing safely, the forward elbow must remain extended as the arms move to swing the club.
*Source:* Adapted and reprinted with permission from Anne Fewell.

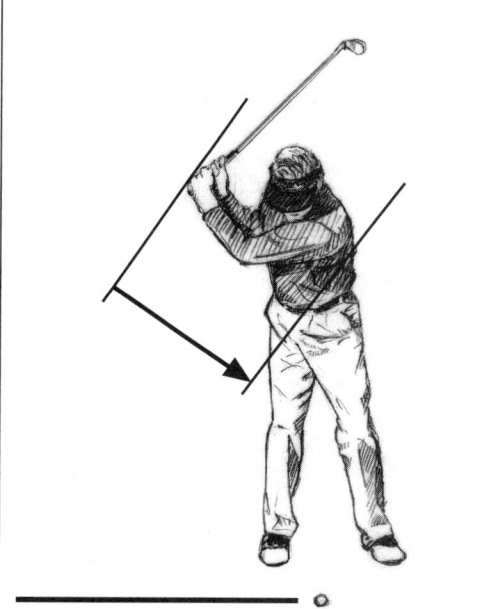

**FIG 1-6** Elbow pain is the second most common complaint from amateur golfers. The collapsed elbow will snap into extension at impact.
*Source:* Adapted and reprinted with permission from Anne Fewell.

## TREATMENT PROTOCOL

**ACUTE:** Apply the RICE protocol to the inflamed tendons:

**R**est

**I**ce

**C**ompression

**E**levation

**CHRONIC:** If necessary, when faced with shots in deep rough or sand traps with deep sand, give up a stroke and take an unplayable lie. Away from golf apply the strength and conditioning protocols for the wrist outlined in Part 3, "Barriers to Performance."

**TREATMENT PROTOCOL** The truth is, many of your patients who play golf are borderline obsessive about playing golf—injured or not. Therefore, effective patient management requires professional in-office treatment and physical therapy. One of the most effective soft-tissue therapy techniques I have discovered is *ART* (see www.ActiveReleaseTechnique.com). Active Release Technique

**FIG 1-7** Hitting a shot from the deep rough is an injury risk for the player that has not conditioned his or her body for golf.

(ART) is a patented, state of the art soft-tissue system/movement based massage technique that treats problems with muscles, tendons, ligaments, fascia, and nerves. ART has been developed, refined, and patented by P. Michael Leahy, DC, CCSP. Dr. Leahy noticed that his patients' symptoms seemed to be related to changes in their soft tissue that could be felt by hand. By observing how muscles, fascia, tendons, ligaments, and nerves responded to different types of work, Dr. Leahy was able to consistently resolve over 90% of his patients' problems. He now teaches and certifies health care providers all over the world to use ART. In addition, there are specific home-care treatment protocols your patients can follow. I have discovered an effective home remedy for golfers with chronic stiffness that includes four key ingredients:

- **Capsaicin:** Synthesized from hot peppers; an FDA-approved over-the-counter agent for pain relief; known to deplete the level of substance P (SP), a neurotransmitter to the brain. Substance P is involved in the transmission of pain impulses from peripheral receptors to the central nervous system. It has been theorized that it plays a part in fibromyalgia. Capsaicin has been shown to reduce the levels of Substance P probably by reducing the number of C-fiber nerves or causing these nerves to be more tolerant. By depleting SP, the pain signal is diminished.

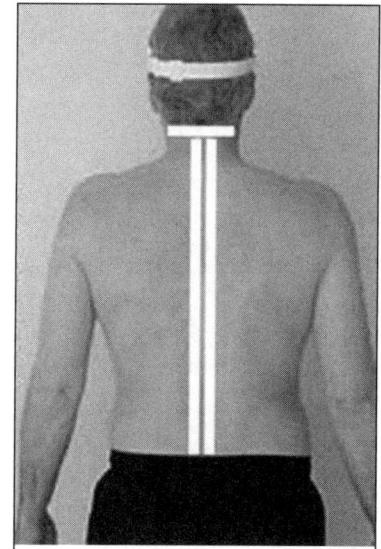

- **Turmeric:** More commonly known as a spice in Indian food; known to reduce inflammation and inhibit the production of inflammatory enzymes.

- **Glucosamine:** A well-known, potent anti-inflammatory.

- **Methylsulfonlmethane (MSM):** A natural form of bioavailable sulfur that acts as a cellular transporter to carry the ingredients efficiently through the dermal layer.

The remedy Play On Lotion can be carried in the golf bag. For office treatment, I recommend supplementing with ultrasound to drive the solution deep into the connective tissues.

**FIG 1-8** Play On Lotion should be applied to the suboccipital and paraspinal muscles 15–20 minutes before playing golf.

# PHYSICAL EXAMINATION

## CAN YOUR PATIENT GET INTO THE RIGHT POSITION?

When you look at photographs of professional golfers swinging the club, you see them in positions that are out of reach for the average golfer. Remember, most professionals started swinging the club as a kid and never stopped. Therefore, they have been able to maintain a youthful flexibility well into their adult years.

*Your patients want to hit long and powerful golf shots.*

Long and powerful shots require lots of club head speed. Club head speed is directly proportional to the width and length of the swing arc. The width and length of the swing arc are directly proportional to patient flexibility (Figures 1-9 and 1-10).

Using the left arm as a hand on an imaginary clock face, consider the following using a driver:

1. Left arm to 9 o'clock =
   85 mph of club headspeed = 200 yards of carry

2. Left arm to 10 o'clock =
   110 mph of club head speed = 225 yards of carry

**FIG 1-9** Wide arc

*Source*: Adapted and reprinted with permission from Anne Fewell.

**FIG 1-10** Narrow arc

*Source*: Adapted and reprinted with permission from Anne Fewell.

3. Left arm to 11 o'clock =
   115 mph of club head speed = 240 yards of carry

4. Left arm to 12 o'clock =
   125 mph of club head speed = 270 yards of carry

When you factor in the ground roll ratios, you can understand how the pros drive the ball between 265 and as many as 300 yards. They have the flexibility that allows them to generate tremendous club head speed.

A swing arc that is wide and long requires your patient to able to flex, extend, side-bend, and rotate. There may be many reasons your patients are not flexible. Perhaps they have bad posture. Perhaps they never stretch. Perhaps they have been injured and there is scar tissue. In any event, any amount of improvement in your

> Tell your patients there are no shortcuts. To hit longer and more powerful golf shots they must be willing to improve their posture and their flexibility.

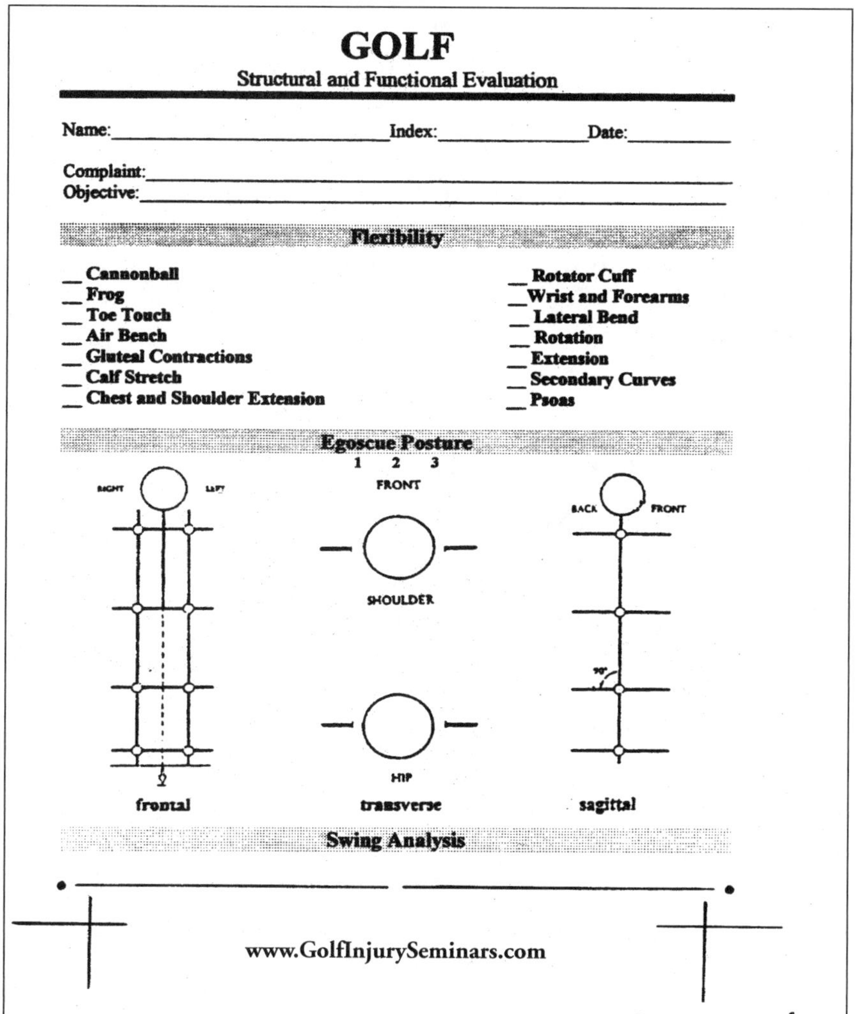

# GOLF
## Structural and Functional Evaluation

Name:_____ Index:_____ Date:_____

Complaint:_____
Objective:_____

### Flexibility

___ Cannonball
___ Frog
___ Toe Touch
___ Air Bench
___ Gluteal Contractions
___ Calf Stretch
___ Chest and Shoulder Extension

___ Rotator Cuff
___ Wrist and Forearms
___ Lateral Bend
___ Rotation
___ Extension
___ Secondary Curves
___ Psoas

### Egoscue Posture

1  2  3

RIGHT      LEFT        FRONT                    BACK      FRONT

SHOULDER

HIP

frontal              transverse              sagittal

### Swing Analysis

www.GolfInjurySeminars.com

**FIG 1-11** Patient exam form

patients' flexibility will improve their ability to hit longer and better-quality golf shots.

The Golf Structural and Functional Evaluation form shown in Figure 1-11 is provided for your review. (See the "Products and Services" section at the end of the book). The following pages describe how to use this form to evaluate your patients' flexibility, posture, and swing.

## INDEX

Your patient's index is a reflection of that person's current level of play. An index can range from 0 to 36. Patients whose index is 0 are playing golf at par. They would be considered highly competent golfers and might have dreams of playing professionally. Patients whose index is 18 means they shoot an average score of 90 on a par 72 course. Patients are often obsessed with their current index, so you need to understand what they are talking about.

## OBJECTIVE

Getting your patients to state objectives clearly for their golf is very important. Some might say, "To be able to play without pain." Some might say, "To be able to hit the ball farther. I'm losing distance." Some might say, "To lower my index." Whatever the objective, once stated, you will be able to help your patients if they will comply with your recommended treatment procedures and training protocols.

# 14-POINT FLEXIBILITY EXAMINATION

When patients describe the experience of tight muscles, they are half right. It's really the three-dimensional web of connective tissue called *fascia* that is tight. Fascial tightness can pull or compress key structures in the body up to 2,000 pounds per square inch. The following sections describe the 14 different clinical examinations that are essential to evaluate any patient who plays golf. You will learn how each examination applies to the golf swing. For any failed examination, you will learn the correction protocol.

When instructing your patients on stretching protocols remind them to
1. Drink lots of water.
2. Stretch slowly.
3. Keep breathing.

## THE CANNONBALL

Basic spinal flexion. Have your patient sit at the edge of a chair or, better yet, squat to the floor. Now tell them to tuck the chin to the chest and wrap the arms around the knees. Hold for 1 minute. Breathe (Figure 1-12).

## THE FROG

An adductor stretch (Figure 1-13). From the seated position, have your patient cross a leg, then push the knee downward. If he or she cannot get the leg parallel to the ground, he or she has failed the test.

The adductors are activated during the transition of weight transfer from the top of the backswing to the start of the downswing. If he or she is too tight, your patient will have trouble pulling his or her extended left arm down the vertical plane and will rotate too quickly into the shot, which generally means he or she will "come over the top."

## TOE TOUCH

Tight hamstrings can prevent the pelvic girdle from normal function. The hips must be in a neutral state for there to be a good secondary curve in the lumbar spine. The secondary curve in the lumbar spine acts as a lever arm to enhance spinal rotation during the backswing. Have your patient bend over and reach to touch the toes (Figure 1-14). Hold for one minute. Breathe.

## AIR BENCH

When your patient swings the club to the top of the backswing, he or she has transferred about 75 percent of the body weight onto the

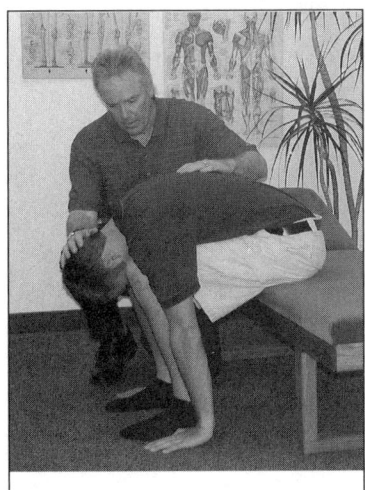

**FIG 1-12** The cannonball

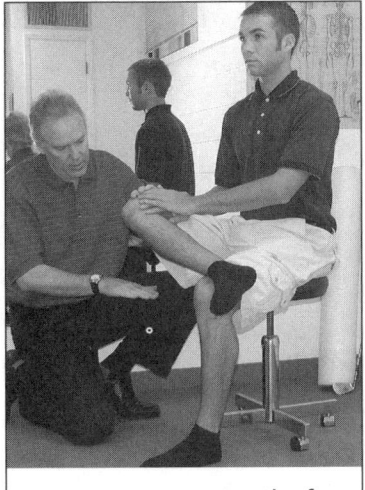

**FIG 1-13** The frog

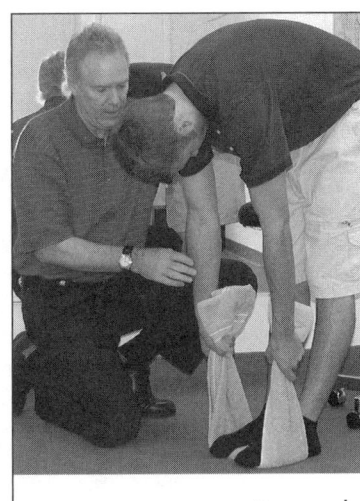

**FIG 1-14** Toe touch

**Note:**
The correction protocol of any failed flexibility test is the test itself. In other words, if your patient experiences painful restriction of the hamstrings, then the patient should stretch the hamstrings everyday.

right quadriceps. This muscle must be fit and functional; otherwise, after five hours of golf and hundreds of swings, legs become fatigued and your patient's golf swing gets "wobbly." Have your patient stand with the back to a wall (Figure 1-15). The heels should be about 12–14 inches from the baseboard. Slowly slide the back down the wall until the thighs are at a 45 degree angle above the knees (do not slide the thighs to parallel to the floor). Once in position, lift the toes up off the floor until the support is from the heels. To be considered baseline functional for golf, patients need to hold this position for 3 minutes. Have your patients start with 1 minute and build to 5 minutes over time.

## GLUTEAL CONTRACTIONS

Have your patient stand and place the fingertips over the gluteal muscles. When the gluteal muscles are tightened, the kneecaps should rotate outward. If your patients cannot rotate their kneecaps outward, they have failed this test. The gluteas maximus is the largest muscle in the body. It acts in concert with the tensor fascia latae to stabilize the hip and knee joints. The knees must be positioned directly over the ankles to provide the hips and shoulders with a stable platform on

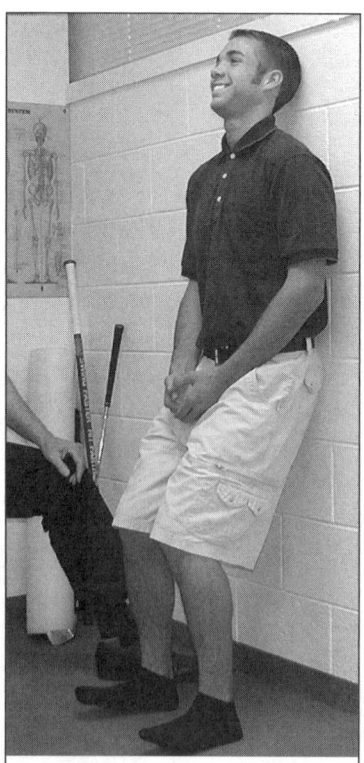

**FIG 1-15** Air bench

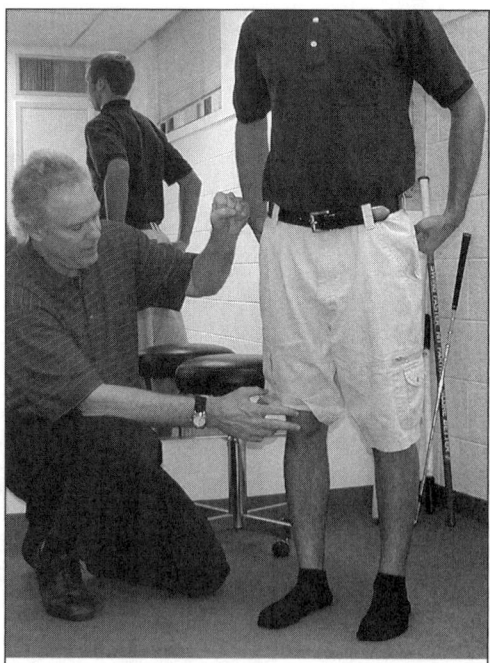

**FIG 1-16** Gluteal contractions

which the torso can rotate. Have your patients strengthen the gluteal muscles by contracting them 30 to 50 times each day. When the gluteals are functional, a strong contraction will torque the femur head and rotate the kneecaps outward (Figure 1-16).

## CALF STRETCH

With your patient in a seated position, gently push the toes toward the knees—first with the leg bent at the knee, then with leg straight. If your patient complains of pain with either maneuver, he or she has failed the test. If the gastrocnemius and the soleus muscles are too tight, your client will have a tendency to move the torso up and down when swinging the club. (Golf professionals refer to this as an inability to maintain a consistent "spine angle.") The feet essentially need to stay planted until the extended forward arm has descended the vertical plane to about 7 o'clock position. Just before impact with the ball, the back heel will come off the ground as the body weight transfers through the shot and into the forward leg (Figures 1-17 and 1-18). This calf stretch is a bilateral stretch. Hold for 1 minute each side. Breathe.

## CHEST AND SHOULDER EXTENSION

Have your patient stand and place the arms behind the back, thumbs touching, palms up. Without bending forward at the waist, he or she should be able to extend the arms at a minimum of

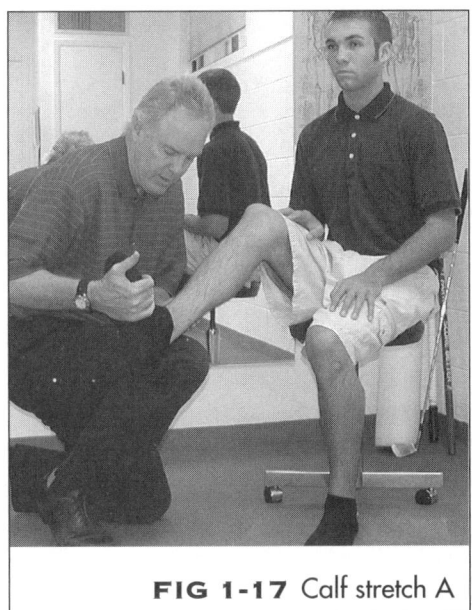
**FIG 1-17** Calf stretch A

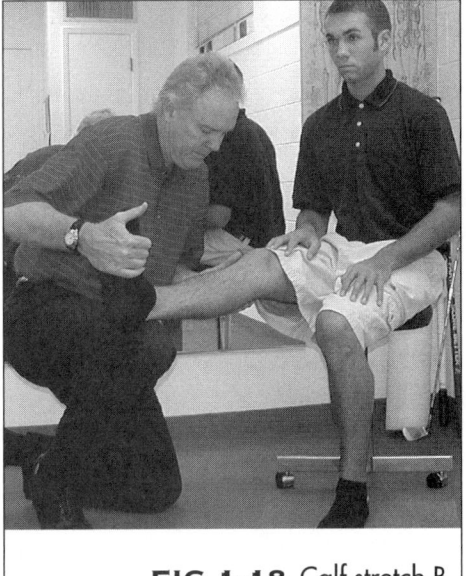
**FIG 1-18** Calf stretch B

30 degrees relative to the upright spine. This is a necessary stretch to release the capsular ligaments surrounding the shoulders. Adhesions in the shoulder girdle cause restriction in the ability of the extended left arm to climb past 9 o'clock on the vertical plane during the backswing (Figure 1-19). Hold for 1 minute. Breathe.

## ROTATOR CUFF

This is another essential stretch for the adhesions in the shoulder girdle. Have your patient stand and extend the arms directly parallel to the ground in front of the torso. Have him or her take the right hand and place it on top of the left wrist. Now try and "sweep" the extended arm across the front of the body. Repeat the same procedure for the other arm. Remind your patient to keep the elbow extended and straight (Figure 1-20). Do not permit rotation of the torso. This is a bilateral stretch. Hold for 1 minute each side. Breathe.

## WRIST AND FOREARM

Have your patient stand. Place the right elbow against the right side and bend the arm at the elbow until the forearm is parallel to the ground. Next, with the left hand, extend the right wrist until the back of the hand is 90 degrees to the forearm. While holding the right wrist in extension, have your patient attempt to extend fully the arm forward away from the body. Repeat this procedure for the left arm.

When performing this test, your patients must be able to get their elbow fully extended while the back of the hand maintains a

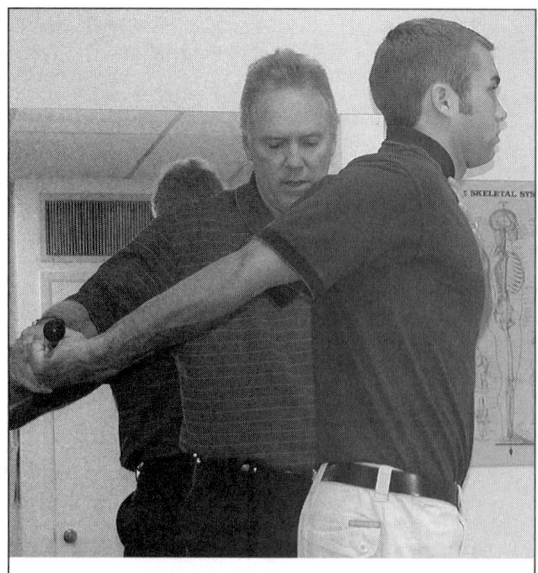

**FIG 1-19** Chest and shoulder extension

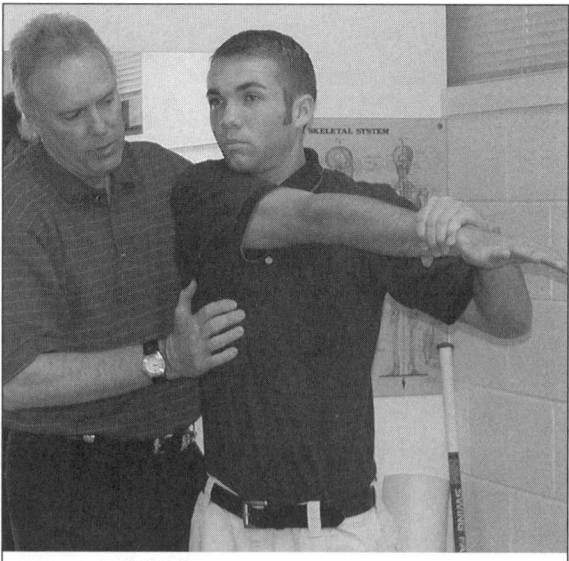

**FIG 1-20** Rotator cuff

90 degree relationship to the forearm. If they cannot do this, they fail this test.

Once your patients learn the importance of setting the correct incline plane with different length clubs, they will understand that their wrists and forearms need to be strong and supple. Check both the flexor and extensor tendons of the forearm (Figures 1-21, 1-22, 1-23, and 1-24). These are bilateral stretches. Hold for 1 minute in each position.

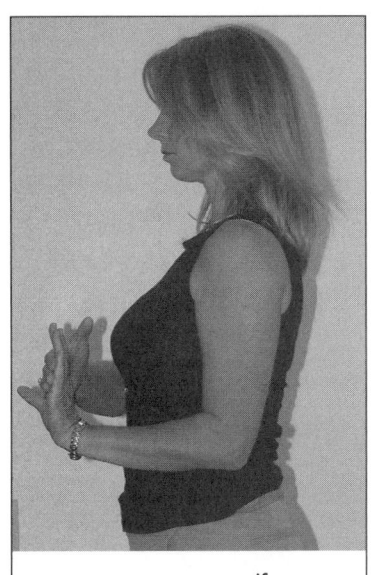

**FIG 1-21** Wrist/forearm stretch for flexor tendons of forearm (A)

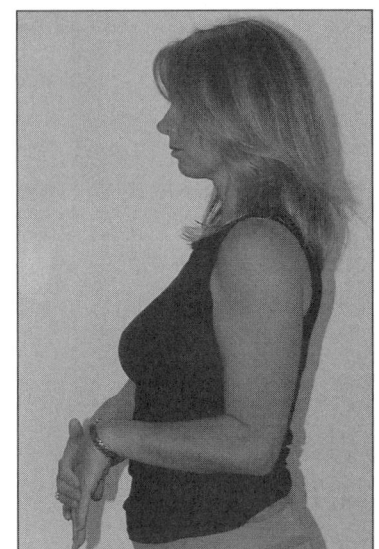

**FIG 1-22** Wrist/forearm stretch for extensor tendons of forearm (A)

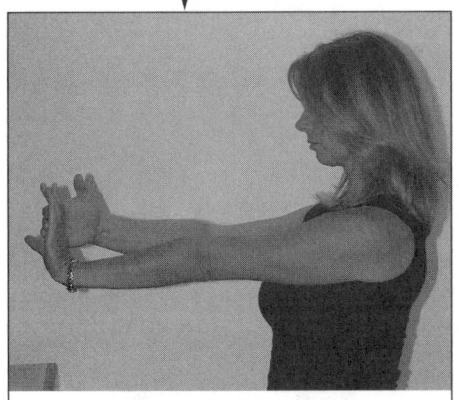

**FIG 1-23** Wrist/forearm stretch for flexor tendons of forearm (B)

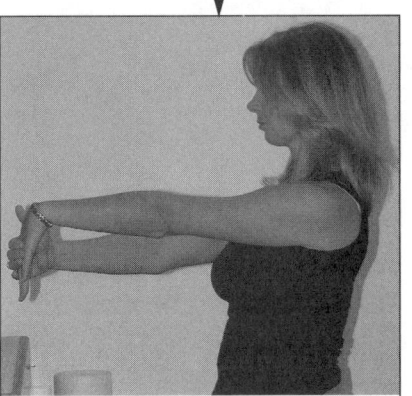

**FIG 1-24** Wrist/forearm stretch for extensor tendons of forearm (B)

## LATERAL BENDING

From a standing position, with the palm of the hands to their sides, have your patients run their hands down the side of their thigh until the fingertips reach the knee—without rotation of the torso. Failure to reach the knee is a failed test. The ability of patients to bend laterally is essential for them to keep their extended left arm parallel to the plane line when swinging the club. If your patients cannot bend laterally, their torso will move up and down when swinging the club. This is a bilateral stretch (Figure 1-25). Hold for 1 minute each side. Breathe.

## ROTATION

Spinal rotation is another essential component in your patients' fitness for golf. Make sure their ability to rotate is equally bilateral. If it is not, they are more prone to injury as a result of loss of bilateral function (Figure 1-26). Hold for 1 minute each side. Breathe.

## EXTENSION

Many of your patients live and work in a sedentary lifestyle. They sit in flexion posture at work. They sit in flexion posture in their cars.

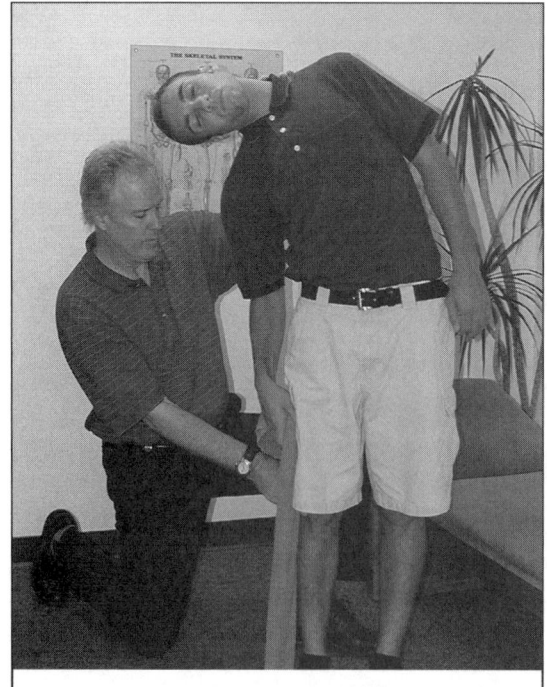

**FIG 1-25** Lateral bending

**FIG 1-26** Rotation

They sit in flexion posture at home. To safely swing a golf club, your patients' spine must have the functional capacity to entend. Extension of the cervical and lumbar spine is necessary in order to maintain balance at the address position. Extension of the lumbar and thoracic spine is necessary to keep the club shaft "on-plane" after impact with the ball. Extension of the entire spine is necessary to hold your finish position. Failure of the spine to extend properly causes your patients to "pull up and out" of the shot. This results in shots hit offline.

1. **Static back:** Knees *must* be positioned directly over hip joints (Figure 1-27). Keep patient down until posterior superior iliac spine pressure into the floor feels equal bilaterally. (5–20 minutes).

2. **Swiss ball:** Make sure your patient's head is always in contact with the ball (Figure 1-28). Hold for 1 minute.

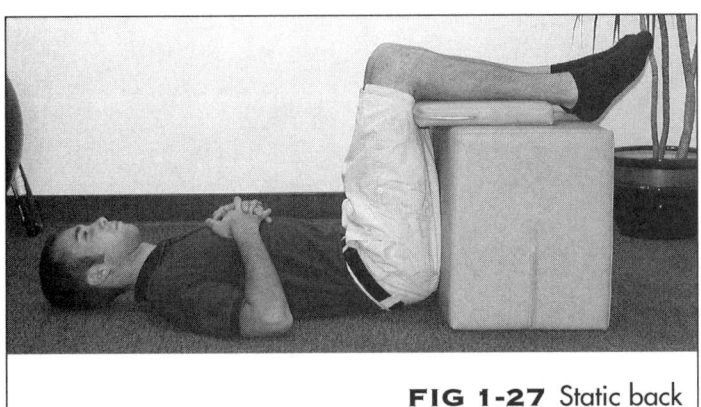

**FIG 1-27** Static back

**FIG 1-28** Swiss ball

3. **Thoracic roller:** Very demanding stretch. Go slow. Always support the head and neck (Figure 1-29). Hold for 20 seconds. Move roller to different spinal segments.

## SECONDARY CURVES

The secondary curves of the spine are essential for mechanical leverage (Figure 1-30). The muscles are stronger when they contract against the curves. In addition, the torso will rotate farther during the backswing. Rolled towels are effective fulcrums for restoration of flattened secondary curves.

Many patients will have trouble resting comfortably while supine on the floor with their knees bent. Before proceeding to rolled towels, make sure your patients can do this for 20 minutes at a time with no discomfort. Be patient, this might take a few weeks.

## TRACTION WITH TOWELS

Use a bath-sized towel for the cervical spine. When rolled it should be about 4 inches in diameter. When properly placed, the back of the head should be able to touch the floor (Figure 1-31).

For the lumbar spine, use a hand towel wrapped in a wash cloth. The diameter should be about 2 inches when rolled. When properly placed, the towel rests just above the belt line at L3. The sacrum and the thoracic spine should remain in contact with the floor.

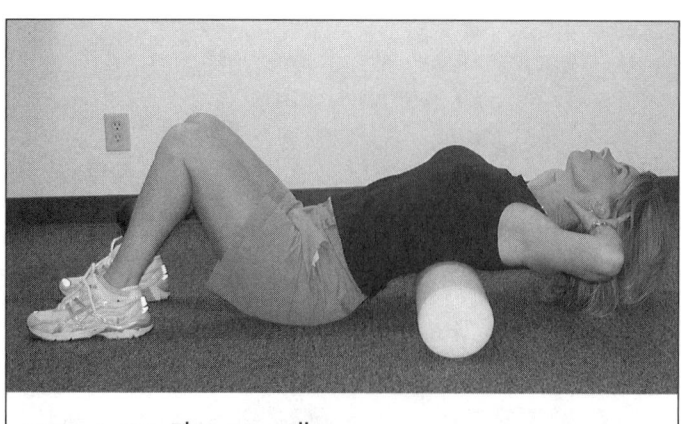

**FIG 1-29** Thoracic roller

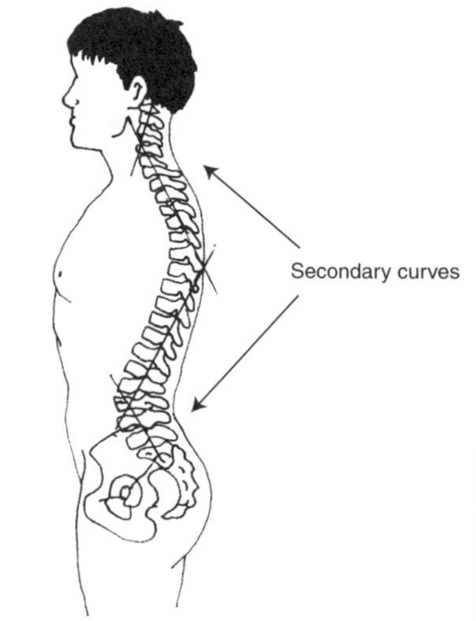

Secondary curves

**FIG 1-30**

Caution your patients that being uncomfortable on the towels for the first week or so is normal. Instruct them to start with 3 minutes a day. They can add more time every other day until they can rest comfortably on the towels for 20 minutes.

## PSOAS

When the iliopsoas is short and tight, the quadriceps are recruited to initiate the flexion of the hip when walking. In addition, tight psoas muscles restrict normal pelvic position and function. With one thigh at 90 degrees to the floor, the other thigh extended with the heel to the floor, have your patient contract the quadriceps of the lengthened leg (Figure 1-32). If the tightest point of contraction is near the knee, the iliopsoas is too tight. Have you patients quietly remain in this position; gravity will do the stretch. Have them contract the lengthened quadriceps every 10 minutes. When the tightest point of contraction is above mid-thigh (closer to the anterior superior iliac spine), the iliopsoas has released. Switch legs and repeat.

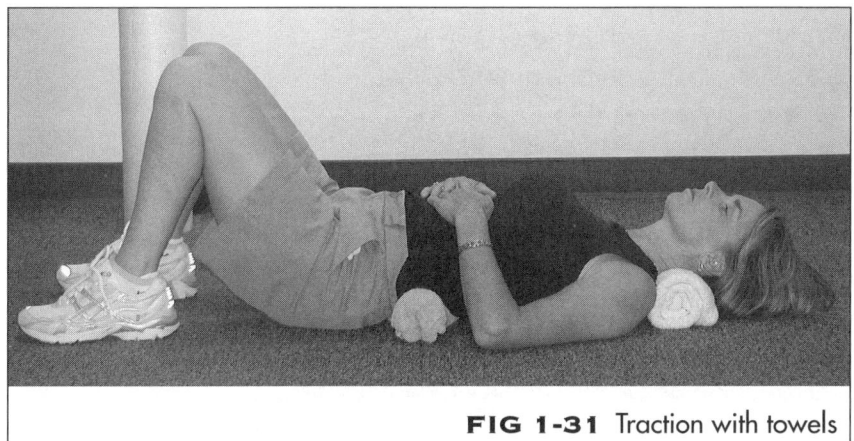

**FIG 1-31** Traction with towels

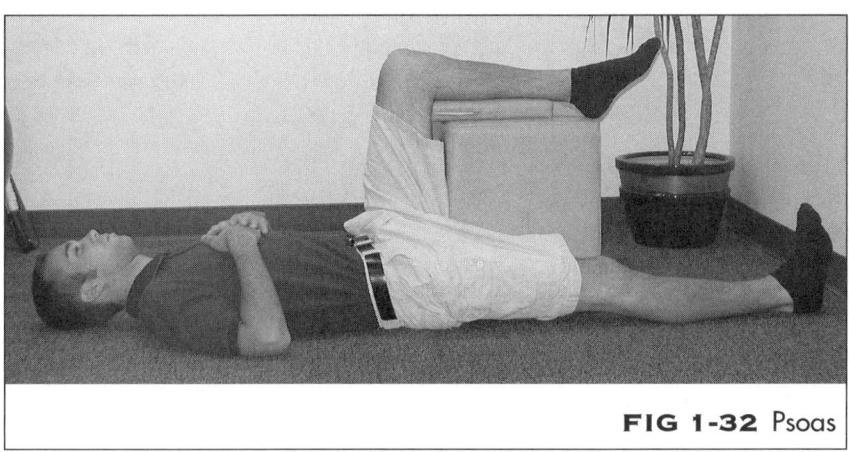

**FIG 1-32** Psoas

# GOLF POSTURE

Pete Egoscue wrote a book called *The Egoscue Method of Health Through Motion* (Harper Collins, 1992), in which he identifies three dysfunctional postural conditions (described in the following box) identified by the position of the pelvic girdle. More important, Egoscue developed correction protocols for each condition. I have found his work to be nothing less than brilliant. If your patients are willing to do their "homework," they can change their posture. Better posture provides the body a better opportunity for bilateral functional movement. The goal is for your patients to recover the function of their inherent bilateral body.

---

### Condition I

The pelvic girdle has tilted forward (Figure 1-33). The general cause is very tight hip flexors.

### Condition II

The pelvic girdle has rotated (Figure 1-34). One ilium is subluxated anterior-superior (AS); the other has compensated posterior-inferior (PI). Once the ilium subluxates PI, the support for the shoulder gives way. The shoulder will drop and cause the torso to rotate.

### Condition III

The pelvic girdle has tilted "under" (Figure 1-35). This condition is generally associated with a complete loss of and sometimes a reverse of the lumbar curve.

---

The correction sequencing for these conditions is beyond the scope of this text. Please refer to Egoscue's book (1999) or visit his Web site (www.egoscue.com). His methods are valuable resources for any patient in your practice.

## PATIENT SWING ANALYSIS

Have your patient use a club to make a few practice swings. You could provide a synthetic grass mat and some foam golf balls if you want, but this is not necessary. If your treatment rooms have an 8-foot ceiling, take an old golf club to a pro shop and have them cut it down and put on a new grip. The length of club, however, is irrelevant to your analysis.

In the analysis, you are going to look for and identify three different components of the patient's golf swing.

"That's it?" Yes, that's it. "Hey, but what about . . . ?" Every single "what about" takes care of itself if your patients train to perfect execution these three key points.

*When you watch your patient swing the club, you must train yourself to look at one component at a time. There are too many moving parts of body and club to try and do otherwise.*

## THREE KEY POINTS

1. Is the position of the back foot perpendicular to the plane line?

2. Does the extended forward arm trace parallel to the plane line during the backswing and the downswing?

3. Does the club shaft point the plane line throughout the swing?

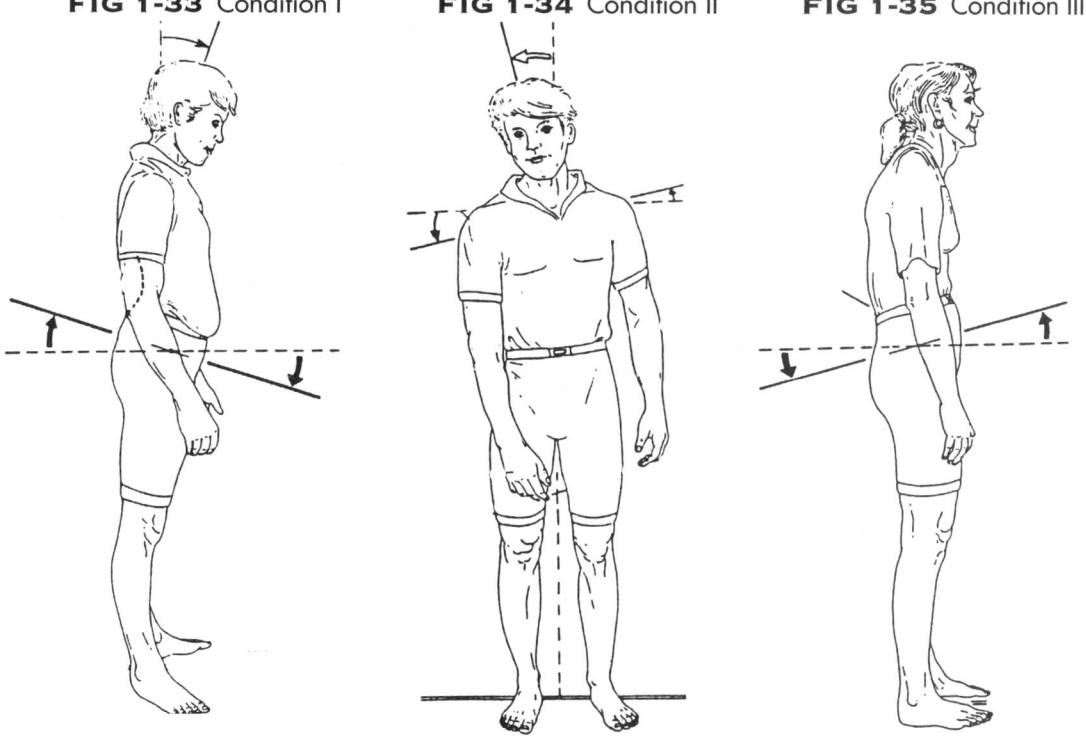

**FIG 1-33** Condition I     **FIG 1-34** Condition II     **FIG 1-35** Condition III

*Source*: Reprinted with permission from *The Escogue Method of Health Through Motion.* ©1992, courtesy of HarperCollins Publishers.

## GOLF POSTURE: THE ATHLETIC POSITION

When the decision is made to hit a golf ball, the club shaft will twirl around the body at anywhere from 90 to 115 mph. If your patient is not in the athletic position at the start of a swing, it's easy for him or her to become unbalanced. Loss of balance puts a patient at risk of injury.

*The secondary spinal curves relate to your patients' ability to rotate for a powerful golf swing.*

For any athlete, the athletic position can be described as a state of readiness for motion and movement in any direction. There is positional alignment of the key load-bearing joints of the body (the shoulders, the hips, the knees, and the ankles) in the frontal and saggital planes. The upright spinal column maintains positional lordosis of the secondary curves.

Sound golf posture is both static and dynamic (Figure 1-36). Static posture is responsible for your patient's overall alignment at the address position before the swing begins. The ability to maintain good posture during the golf swing is a measure of a golfer's dynamic postural strength and fitness.

We have been taught that the position of the structure dictates the function of the structure. Good golf posture requires the secondary spinal curves to be in lordosis (Figure 1-37). The secondary spinal curves provide mechanical leverage for strength and facilitate the patient's ability to rotate. Poor golf posture with a collapse of the secondary curves restricts spinal motion. This causes the golfer

**FIG 1-36** The athletic position at address. The feet, knees, hips, and shoulders are parallel to the plane line.

*Source*: Adapted and reprinted with permission from Anne Fewell.

**FIG 1-37** The correct golf posture. The lordosis of the cervical and lumbar secondary curves.

to compensate by overswinging. Continuous overswinging leads to soft-tissue injuries.

Remember, the golf club swings around the body of your patient at about 90 miles per hour. When the club is moving out of position, your patients are at risk of injury.

There are three key points relative to golf swing mechanics you must learn to identify (Figures 1-38, 1-39, and 1-40). These key points are addressed in detail in Chapter 2, "Swing Mechanics."

You are a health care professional, not a golf professional. Nevertheless, it is your responsibility to determine whether your patients are swinging the club safely.

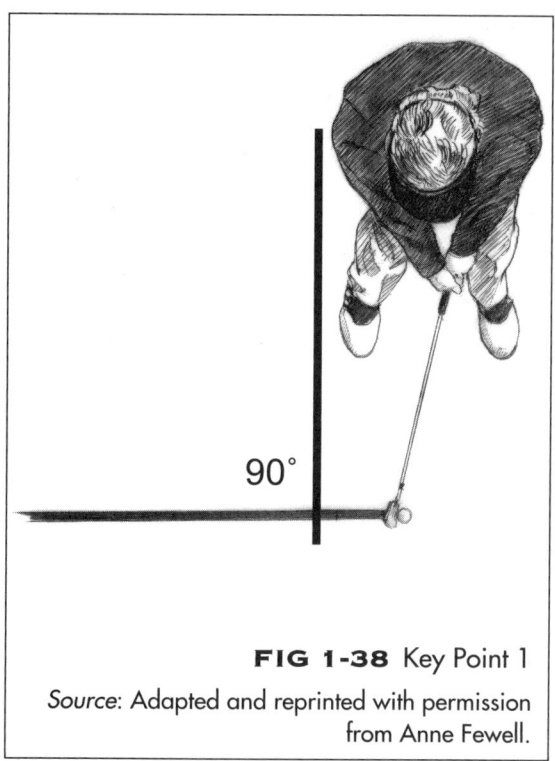

**FIG 1-38** Key Point 1

*Source*: Adapted and reprinted with permission from Anne Fewell.

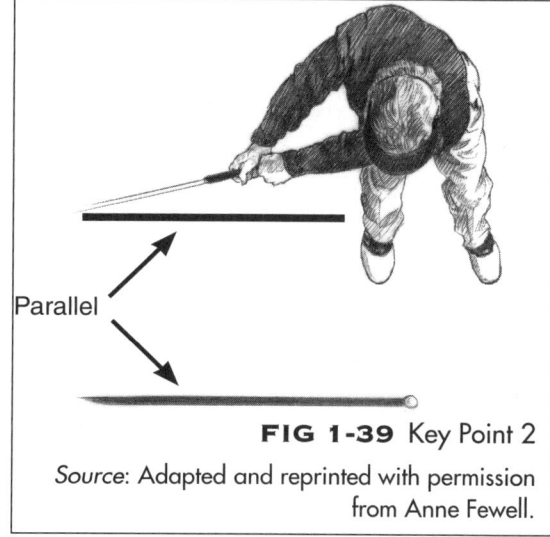

**FIG 1-39** Key Point 2

*Source*: Adapted and reprinted with permission from Anne Fewell.

**FIG 1-40** Key Point 3

*Source*: Adapted and reprinted with permission from Anne Fewell.

# Chapter 2

## SWING MECHANICS

### Three Key Points with Correction Protocols

# INTRODUCTION

## ARE YOU IN THE RIGHT POSITION?

Get a golf club, swing it to the top of your backswing, and stop. Hold it there in the correction position. Feel free to make a few adjustments. Just make sure you are in the right position.

> The truth is, you won't know if you're in the right position until after you've hit the ball.

> I quickly realized that without the immediate feedback of the flight path of the ball, I had no idea if I was swinging correctly.

If you're ready to be perfectly honest about the state of your golf game, you will confess you don't have any idea if you are in the right position. I know you have some ideas, but I also know you have no objective proof.

Consider the position of your forward arm. Have you moved it too far behind your body? Is it too far in front of your body? You remember being told, "Take it straight back." Was that the forward arm or the club head? Straight back relative to what?

Consider the club head. Before you swing, the club head is in front of your body. At the top of the swing, it's behind your body. Then, on the downswing, it's moving on an angle somewhere between vertical and horizontal. Well, there are lots of angles between vertical and horizontal. You remember reading how important it is to "swing on-plane." Where exactly is "on-plane"?

In 1999, I was hitting a 7-iron into the green during a tournament when I separated a muscle from the rib cage under my forward arm. For the next 3 months, I could not hit a blade of grass, much less a golf ball, without severe pain. During the healing process, I thought I'd keep swinging the club through the air to stimulate the motor memory I had worked so hard to groove.

My background before golf was the study of human biomechanics. I had 20 years of experience as a doctor of chiropractic in the study, diagnosis, and treatment of neck

**FIG 2-1** Dr. Jeff Blanchard

and back pain caused by faulty biomechanics of muscles, bones, joints, ligaments, and tendons.

In medicine, doctors can always compare the examination findings of the patient with relative normal values. For example, I would say, "Mr. Jones, you should be able to rotate your chin 70 degrees to the right. You can rotate only 35 degrees. You have a 50 percent loss of normal motion in your neck. Let's find out why."

In the physical sciences, all available data are measured. Measurement is mathematics. Mathematics is the universal language. The number 4 in the United States is the same number 4 in Germany and Japan.

When the golf swing begins, the current model of instruction is without objective points of reference for when you are in or out of position. Therefore, golfers swing every which way and never the same way twice.

Imagine knowing whether you are in or out of position before you ever hit the ball. My imagination led me to tape penlights to both ends of the club shaft and to swing the club in my darkened garage. Here is what I discovered: When you swing the club, there is an exact path of movement of the forward arm and an exact path

Measurement begins when an origin is established as a point of reference (Figure 2-2).

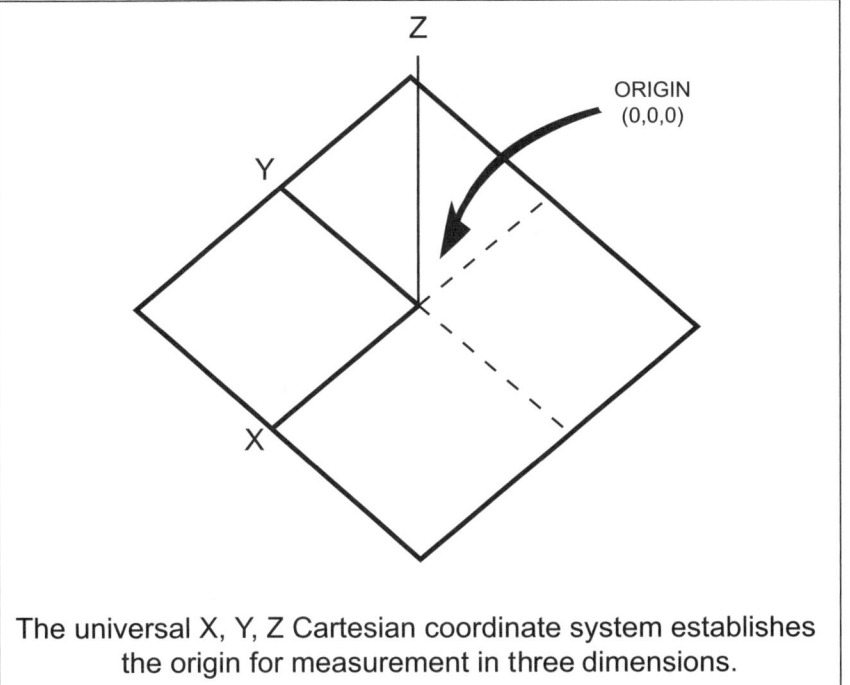

The universal X, Y, Z Cartesian coordinate system establishes the origin for measurement in three dimensions.

**FIG 2-2**

*Source*: Adapted and reprinted with permission from Anne Fewell.

**Plane line:**

From the golf ball to 10 feet behind the golf ball (Figure 2-3).

of movement of the club shaft. The movement is measurable relative to reference points.

Simply stated, use the Plane Line as your point of reference (Figure 2-3).

A. Your back foot must be perpendicular to the Plane Line (Figure 2-4).

B. Your forward arm must swing parallel to the Plane Line (Figure 2-5).

C. The shaft of the club must always point to the Plane Line (Figure 2-6).

Now that an origin has been established for the movement of your forward arm and the position of the club shaft during the golf swing, you can know immediately when you are in or out of position before you ever hit the ball.

You can use the lines in Figure 2-7 to train yourself to swing correctly by tracing. The position of your feet, the movement of your

Blanchard's Plane Line
begins at the golf ball and then extends 10 feet behind the golf ball

**FIG 2-3**

*Source:* Adapted and reprinted with permission from Anne Fewell.

forward arm, and the club shaft can now be objectively measured relative to points of origin to determine whether you are out of position before you hit the ball.

A. Back foot is perpendicular to the Plane Line.

B. Forward arm swings parallel to the Plane Line.

C. Club shaft always points to the Plane Line.

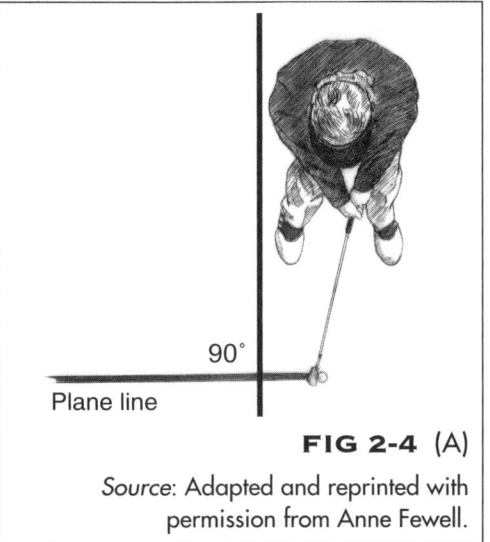

**FIG 2-4** (A)

*Source:* Adapted and reprinted with permission from Anne Fewell.

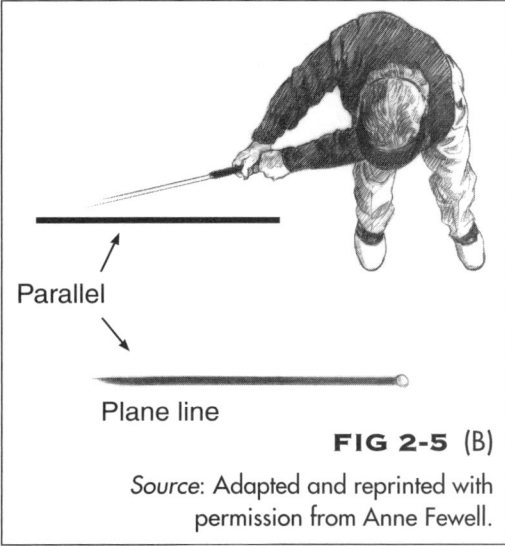

**FIG 2-5** (B)

*Source:* Adapted and reprinted with permission from Anne Fewell.

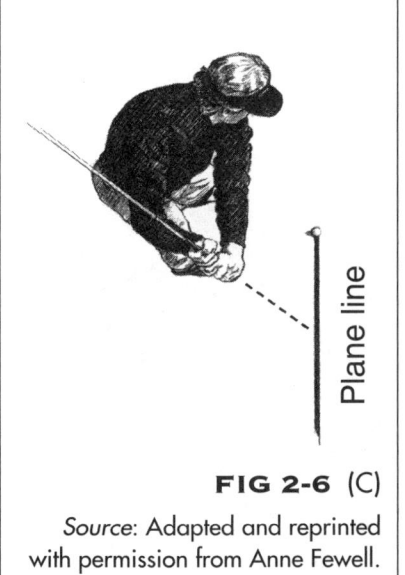

**FIG 2-6** (C)

*Source:* Adapted and reprinted with permission from Anne Fewell.

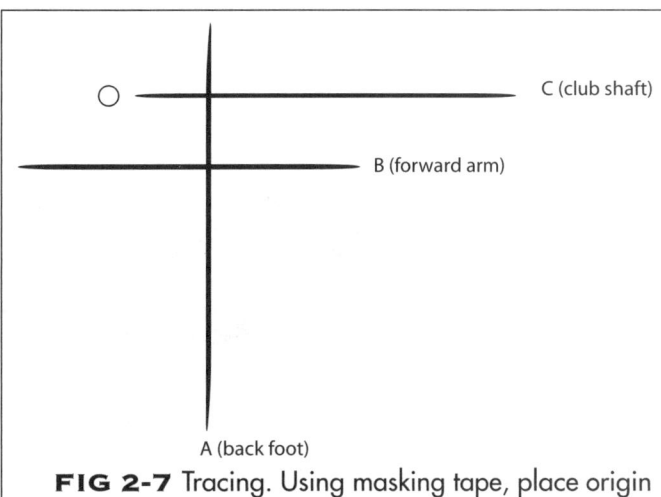

**FIG 2-7** Tracing. Using masking tape, place origin lines A, B, and C on the floor. When moving your body or the club shaft, you will know the moment you are out of position before you ever hit the ball.

*Source:* Adapted and reprinted with permission from Anne Fewell.

# KEY WORDS AND DEFINITIONS

Effective communication stops the moment you hear a word you do not understand. To avoid embarrassment, you might pretend to understand what was said, but in reality you have stopped listening as your mind wanders as you wonder what that word meant.

There are 10 key words and definitions you will read throughout this text and hear listening to the CD. Take a moment to understand their meaning. I hope these words become the new language of golf.

1.  **Geometry:** The measurement and relationships of points, lines, angles, and surfaces. In golf, we use geometry to establish objective points of reference for the movement of your body and the club shaft during the golf swing.

2.  **Training:** To prepare physically, as with a regimen. In golf, let's agree that to play better golf it takes time. You don't take a lesson to climb a mountain; you train to climb a mountain.

3.  **Training zone:** Definition: From 6 o'clock to 9 o'clock on an imaginary clock face (Figures 2-8 and 2-9). When you read that the forward arm must stay parallel to the Plane Line or the club shaft must point at the Plane Line, it means inside the training zone. If you can control the

**FIG 2-8** Training zone

movement of your body and club shaft inside the training zone, you will improve.

4. **Blanchard's Plane Line:** From the target, the Plane Line begins at the golf ball and extends to 10 feet behind the ball.

## 10 feet

5. **Horizontal plane:** Level to the ground. The top of a table rests on the horizontal plane. A merry-go-round rotates on the horizontal plane.

6. **Incline plane:** On an angle. The back of a chaise lounge by the pool is on an incline plane. A slide in the park is built on an incline plane.

**FIG 2-9** Training zone

7. **Parallel:** Two lines that never get closer together or farther apart. The double yellow lines in the street are painted parallel to each other. Railroad tracks are built parallel to each other.

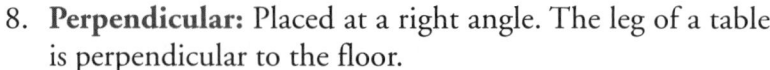

8. **Perpendicular:** Placed at a right angle. The leg of a table is perpendicular to the floor.

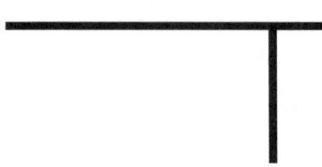

9. **Pivot:** A shaft about which related parts rotate. In golf, the shaft is the right leg. The related parts are the hips and shoulders.

10. **Vertical plane:** Straight up and down. A wall is an example of a structure built on the vertical plane. A ferris wheel rotates on the vertical plane.

# BLANCHARD'S PLANE LINE

## KEY POINTS

The Plane Line represents the origin for alignment of your body.

The Plane Line represents the origin for movement of the forward arm during the golf swing.

The club shaft must always point to the Plane Line.

You have picked out your target and decided on the trajectory and shape of your golf shot. You have calculated the necessary carry to roll ratios against the conditions of the day's weather. You have selected the correct club to execute your intention. You are standing motionless over the ball. You can hear the quiet as you get ready to pull the trigger. Unfortunately, your mind is anything but quiet because you know the truth. You have no idea where the ball is about to go. Now you're feeling the stares of everyone who is waiting. You can feel the panic rising inside because sooner than later, you know you have to swing the club!

As a solution I offer the Plane Line. The Plane Line is a constant point of reference for the alignment of your body before you swing. The Plane Line is a constant point of reference for the movement of your forward arm during the swing. The Plane Line is a constant point of reference for the angle of the club shaft during the swing.

> A. Be sure your back foot is perpendicular to The Plane Line. (Figure 2-12).

The position of your right foot will dictate the direction of the flight path of the ball. In addition, when you square up your right foot to the Plane Line, your hips and shoulders will coil quickly

*Golf is difficult because we so easily lose our way.*

*Training with the Plane Line gives you immediate feedback so you can know whether your body and the club are in the correct positions before you ever hit the shot.*

Plane line

10'

**FIG 2-10**

*Source*: Adapted and reprinted with permission from Anne Fewell.

during the backswing. A strong, tight coil *with your weight inside the back foot* is a necessary step to generate kinetic energy for a powerful downswing.

Before you swing the club, be sure that your hips and shoulders are parallel to the Plane Line (Figure 2-11).

This will ensure proper alignment of your body before you swing the club. Well over 80 percent of players are set up closed to the Plane Line at address. Before they ever swing the club, the ball is already going to the right of their intended target.

When swinging the club:

B. Keep your forward arm parallel to the Plane Line (Figure 2-13).

C. Always point the club shaft at the Plane Line (Figure 2-14).

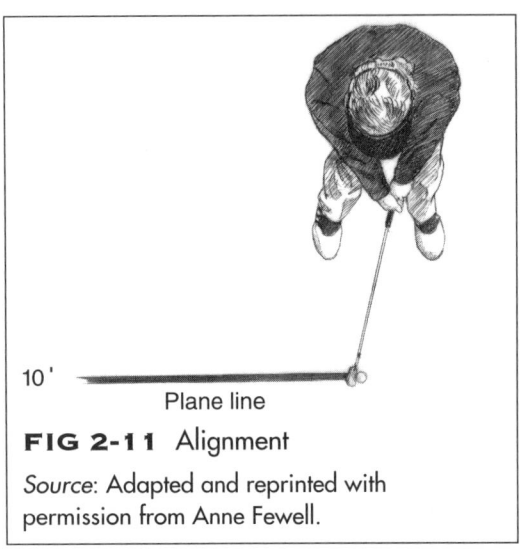

**FIG 2-11** Alignment

*Source*: Adapted and reprinted with permission from Anne Fewell.

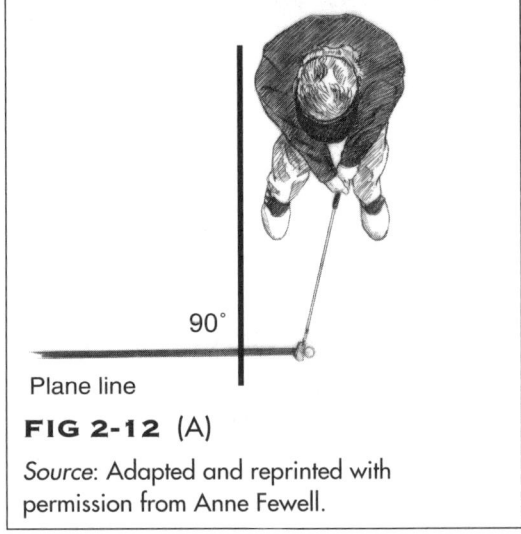

**FIG 2-12** (A)

*Source*: Adapted and reprinted with permission from Anne Fewell.

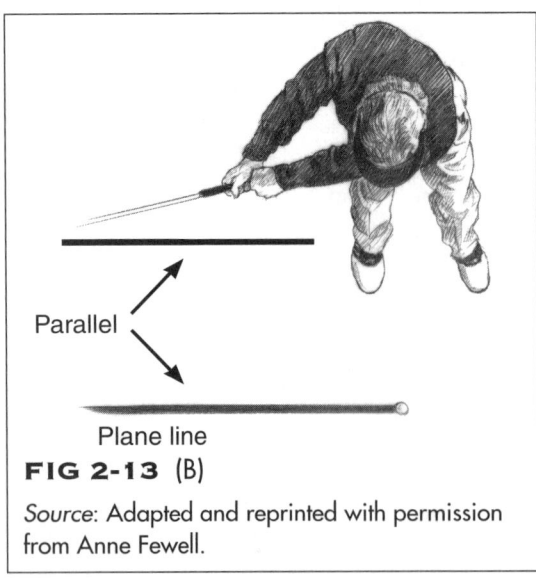

**FIG 2-13** (B)

*Source*: Adapted and reprinted with permission from Anne Fewell.

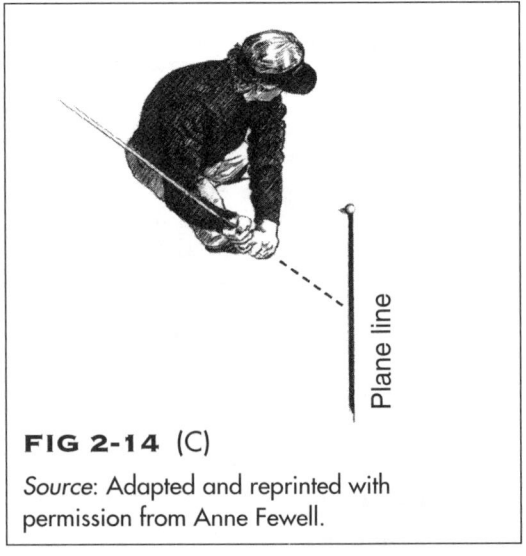

**FIG 2-14** (C)

*Source*: Adapted and reprinted with permission from Anne Fewell.

With the forward arm parallel to the Plane Line you can coil and then release your stored kinetic energy directly toward your intended target. When the club shaft points at the Plane Line, you will be on-plane. In addition, you are assured of striking the golf ball from the inside out.

## THE PIVOT

### KEY POINTS

Before swinging the club, the back foot must be perpendicular to the Plane Line.

During the backswing, your weight must stay coiled inside your back foot.

Never try to rotate your body when swinging the club.

Rotation of the hips and shoulders occurs as a reflex movement necessary to maintain balance when swinging the club in reference to the Plane Line.

The back foot is perpendicular to the Plane Line.

The hips rotate around the stationary back leg.

The shoulders rotate around the stationary back leg.

The weight is coiled inside the back foot (Figure 2-16).

**Definition:**

A shaft about which related parts rotate. In golf, the shaft is the right leg for right-handed players. The related parts are the hips and shoulders (Figure 2-15).

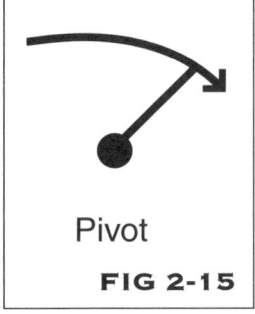

Pivot

**FIG 2-15**

**FIG 2-16**

Your preparation to pivot is the starting place for actually swinging the club. The pivot is essential for building kinetic energy (power) into your golf swing. Kinetic energy translates to club head speed. Club head speed is necessary to compress the golf ball. The more you compress a golf ball, the farther it goes.

**FIG 2-17**

**FIG 2-18** Wide arc

*Source*: Adapted and reprinted with permission from Anne Fewell.

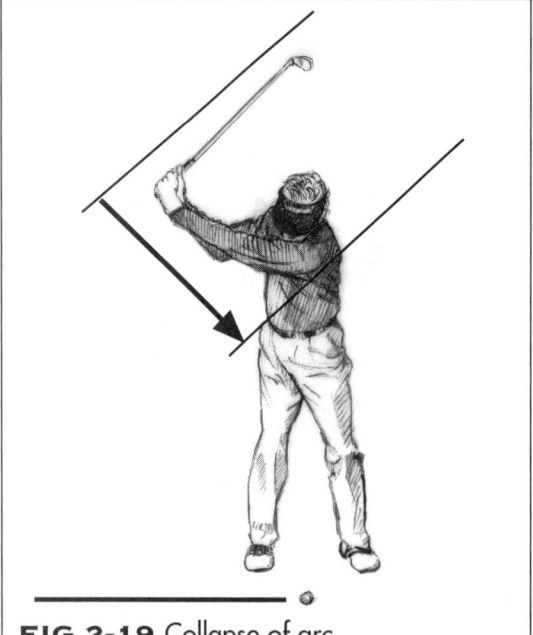

**FIG 2-19** Collapse of arc

*Source*: Adapted and reprinted with permission from Anne Fewell.

The average amateur swings the driver about 90 mph. The ball then flies about 200 yards. The average pro swings the driver about 115 mph. The ball flies about 240 yards. Tiger Woods has been clocked at over 130 mph, and the ball flies well over 270 yards. When you factor in the ground ball, you can understand how the pros drive the ball close to 300 yards.

The right foot for the right-handed player must be positioned perpendicular to the intended target. This is the same as positioning your right foot perpendicular to the Plane Line. During the backswing, the hips and shoulders must rotate to keep your weight inside your right foot.

When you first position your back foot perpendicular to the Plane Line, you will feel "closed" or "pigeon-toed." You will notice that your backswing feels shorter. You might feel pain in your hip and/or lower back. Not to worry. The cause is lack of necessary flexibility, which can be improved by stretching. It is a mistake to turn your back outward at address over the ball to avoid the pain or tightness in your hips and/or lower back.

*Good flexibility permits the hands to extend from the chest. Extension will create a wide arc for power.*

When swinging the golf club, always keep your forward arm parallel to the Plane Line. Eighty percent of the players I have examined before training cannot maintain an extended forward arm past about the 9 o'clock position on an imaginary clock while parallel to the Plane Line. The restriction is a result of loss of joint motion and mobility in the scapula, shoulder girdle, and thoracic spine.

*Poor flexibility causes the forward elbow to bend. The bent forward elbow collapses the arc and diffuses your power.*

You can make dramatic improvements in your game by swinging the club within the limitations of your present flexibility. To keep your forward arm parallel to the Plane Line you might have to shorten your backswing. Until your flexibility improves, use more club. You might try hitting your 6-iron instead of your usual 7-iron from the 150-yard marker.

You have the feeling that you have not taken a big enough backswing, so you cheat. You turn your back foot outward or you bend your forward elbow. At that moment, you immediately diffuse the buildup of kinetic energy and lose power. In addition, your body position and the club shaft are now off-line before the downswing. The ball is likely to go anywhere.

Have you ever wondered why the pitcher's rubber is placed parallel to the leading edge of home plate? It is so pitchers can place their back foot perpendicular to the target. During the windup, a pitcher will pivot the hips and shoulders around the back leg. The pitcher keeps his or her weight coiled inside the back foot, and then releases the stored kinetic energy with a push off the rubber.

A. When the back foot is perpendicular to the target, the hips and shoulders coil rapidly, generating kinetic energy (power) (Figures 2-20 and 2-21).

B. The open back foot creates loss of kinetic power and golf shots hit off-line.

If the pitcher's rubber was incorrectly placed in an open position and pointing to the right of home plate, then that's exactly where the ball would fly when released unless the pitcher made a last-minute correction during release.

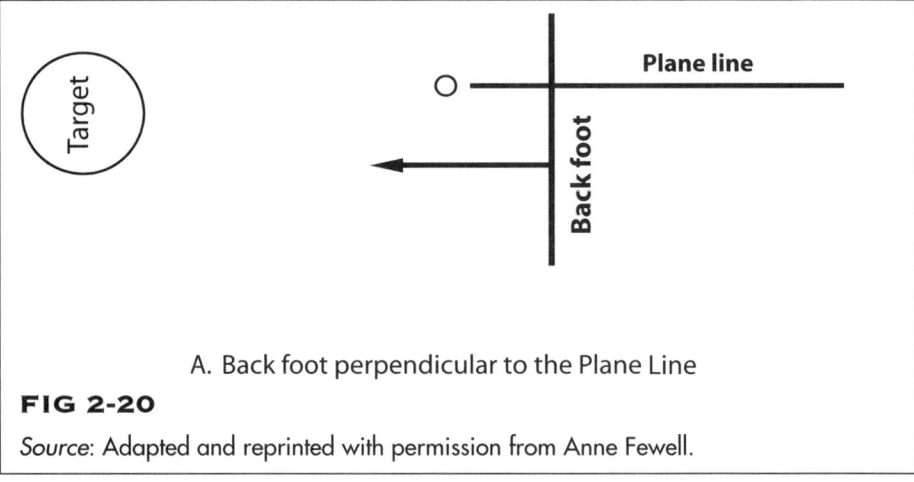

A. Back foot perpendicular to the Plane Line

**FIG 2-20**

*Source:* Adapted and reprinted with permission from Anne Fewell.

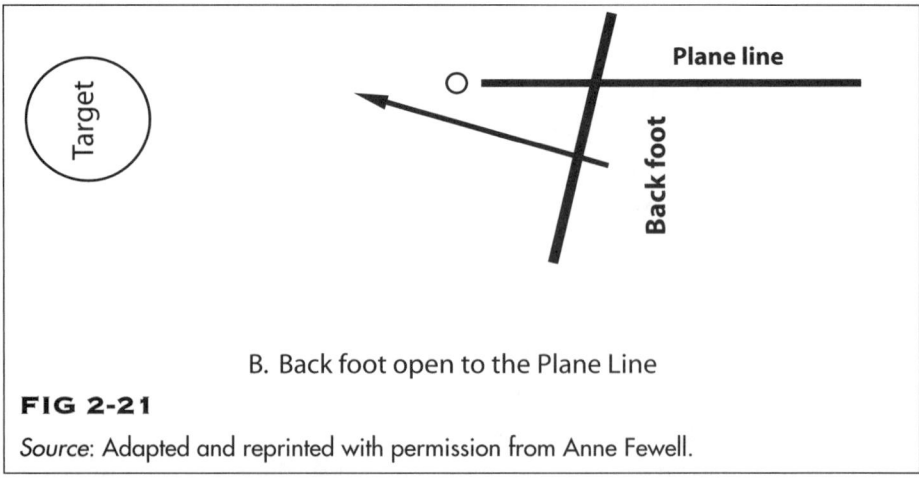

B. Back foot open to the Plane Line

**FIG 2-21**

*Source:* Adapted and reprinted with permission from Anne Fewell.

The hips and shoulders rotate by reflex to maintain balance during the golf swing. The intent of the pivot is to coil as much kinetic energy as possible inside your right foot.

If your shots are mostly going to the right, check the position of your back foot.

> To maximize the buildup of kinetic energy, the extended forward arm must stay out in front of your body and parallel to the Plane Line during the backswing (Figures 2-22 and 2-23).
>
> To ensure the directional release of the kinetic energy, the extended forward arm must stay parallel to the Plane Line on the downswing.
>
> If the forward arm pulls inside the Plane Line during the backswing, the forward arm must push outward away from your body on the downswing.

When the forward arm pushes outward from the body during the downswing, you will "cut across" the ball as your body weight transfers to your left side. This results, more often than not, in hitting a weak, fading slider out to the right of your intended target.

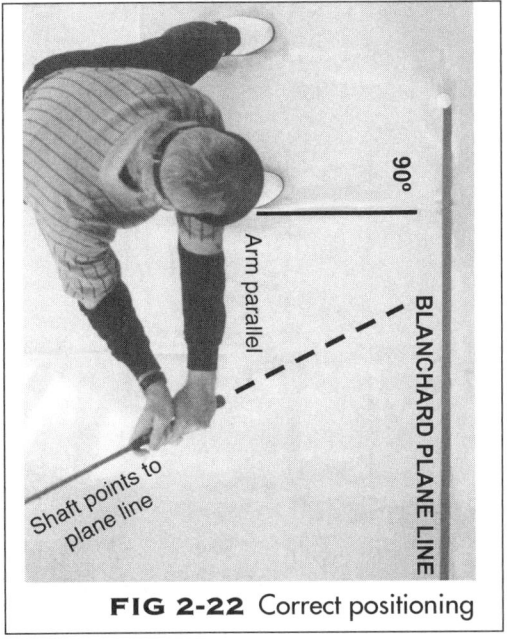

**FIG 2-22** Correct positioning

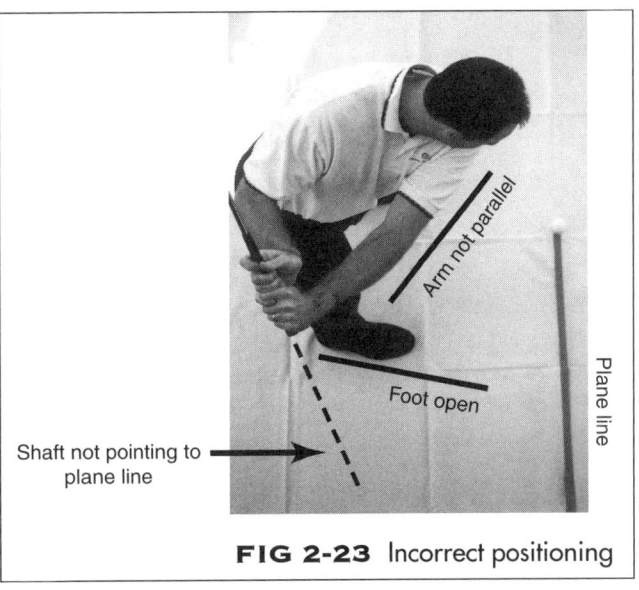

**FIG 2-23** Incorrect positioning

# The Vertical Plane

## KEY POINTS

During the backswing, your forward arm should climb the vertical plane inside the training zone.

During the downswing, your forward arm should descend the vertical plane (Figures 2-24 and 2-25).

Inside the training zone, your forward arm must stay parallel to the Plane Line.

If there were one cornerstone in the golf swing, it would be the path of movement of the extended forward arm. The pivot, the weight transfer, and the rotation of the hips and shoulders are secondary to the direction of movement of your extended forward arm (Figures 2-26 through 2-29).

**FIG 2-24** Backswing

*Source*: Adapted and reprinted with permission from Anne Fewell.

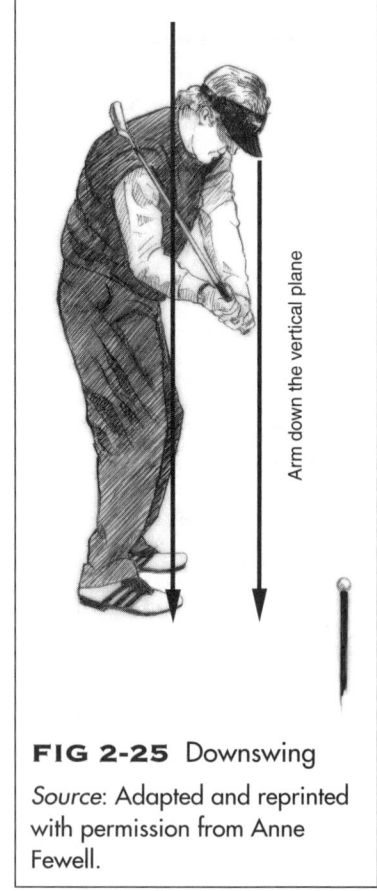

**FIG 2-25** Downswing

*Source*: Adapted and reprinted with permission from Anne Fewell.

**FIG 2-26** Address

*Source*: Adapted and reprinted with permission from Anne Fewell.

**FIG 2-27** Backswing

*Source*: Adapted and reprinted with permission from Anne Fewell.

To stay parallel to the Plane Line, your forward arm must track up and down the vertical plane inside the training zone.

**FIG 2-28** Set point

*Source*: Adapted and reprinted with permission from Anne Fewell.

**FIG 2-29** Downswing

*Source*: Adapted and reprinted with permission from Anne Fewell.

So always remember these points:

> The forward arm is pulled down the vertical plane (Figure 2-30).
>
> The forward arm is parallel to the Plane Line.
>
> The club head is rotating around the body.
>
> The club head will strike the ball from the inside out.

**Definition:**

*On an angle. The back of a chaise lounge is on an incline plane. The slide in the park is built on an incline plane.*

# THE INCLINE PLANE

## KEY POINTS

The shaft of the club is always on an incline plane (Figure 2-31).

The shaft of the club always points to the Plane Line (Figure 2-32).

There are 14 different incline planes, one for every club in your bag. The putter is the 14th club, and generally never swings, though it does have its own incline plane.

If the shaft is on an incline plane during the downswing, the club head will strike the ball from the inside out.

The swing plane may well be golf's most confusing and misunderstood concept. The concept of the swing plane implies a single plane. In reality, the forward arm and the club shaft travel in two different planes when you swing the club.

1. The forward arm pulls down the vertical plane and parallel to the Plane Line toward the intended target (Figures 2-30, 2-33 & 2-34).

2. The club shaft rotates around the body on an incline plane.

**FIG 2-30**

There are two planes, not one. They are separate. Yet, when held together in their respective positions, they create magic. The magic results in that amazing experience when you've struck that pure golf shot that no words could ever describe.

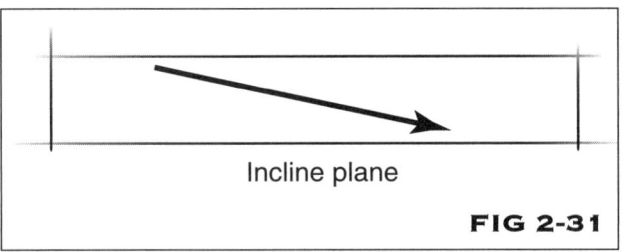

Incline plane

**FIG 2-31**

**FIG 2-33** 1. Forward arm down and parallel to Plane Line

*Source*: Adapted and reprinted with permission from Anne Fewell.

Incline plane

**FIG 2-32**

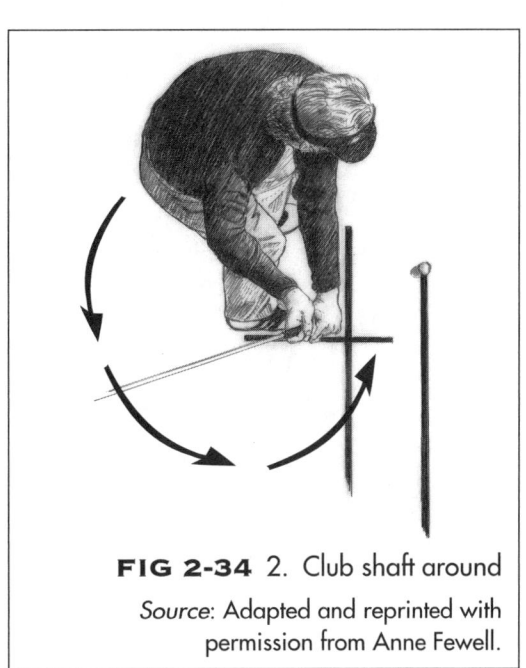

**FIG 2-34** 2. Club shaft around

*Source*: Adapted and reprinted with permission from Anne Fewell.

In baseball, players learn to swing level to the ground (Figure 2-35).

Have you ever wondered why it's easier to hit a baseball moving 90 mph with a bat than it is to hit a stationary golf ball to an intended target? Well, the "swing plane" for the baseball player is easy to find. The reference point is the height of the ball to the playing field (a horizontal plane). As the baseball approaches, hitters adjust their stride and therefore the height of their torso to swing level.

Even a five-year-old understands swinging level. Level is parallel to the ground. The five-year-old does not understand the word *parallel*, but knows that if you swing up at the ball or swing down at the ball, you will likely miss.

In golf, the club swings on an incline plane.

In golf, this plane is very hard to find without an origin for reference. This problem is compounded because golf clubs are different lengths (driver through sand wedge) and golfers are different heights (Figure 2-36).

**FIG 2-35** Horizontal plane
*Source:* © Richard Paul Kane/ShutterStock, Inc.

**FIG 2-36** Incline plane

Before you ever swing the club, note that the shaft of the club rests on an incline plane. The incline plane for a sand wedge is very steep because the club head is only about a quarter step in front of your body. By comparison, the incline plane for the driver is less steep because the club head is closer to a full step in front of your body.

Then, all at once, the club head is behind your body. Between your setup at address, backswing, and downswing, it is very easy for the shaft of the club to get out of position. It's no wonder—you have 13 different full swing clubs and each club has a different incline plane at address. To make matters worse, you have no origin or point of reference to let you know if you are holding the correct incline plane.

I became so frustrated with the problem of identifying the incline for different clubs that I began taping penlight flashlights to the club shaft near the club head and inside the butt end of the grip.

Whether in front of your body or behind your body, the club shaft must stay on plane (Figures 2-37 & 2-38).

The lightbeams on the Swing Light Trainer had the visual effect of extending the shaft of the club.

**DRIVER**
**Shaft in front of body**

45°

**FIG 2-37**

**DRIVER**
**Shaft behind body**

45°

**FIG 2-38**

For more information about the Swing Light Trainer (Figure 2-41), see *www.GolfInjurySeminars.com* or call (805) 772-8298.

Finally, I could answer the question that had been bothering me ever since I started playing golf: "Am I in the right position?" Before using the light trainer with the Plane Line, I never knew for sure until after I hit the ball.

DRIVER
Shaft behind body

FIG 2-39

SAND WEDGE
Shaft behind body

FIG 2-40

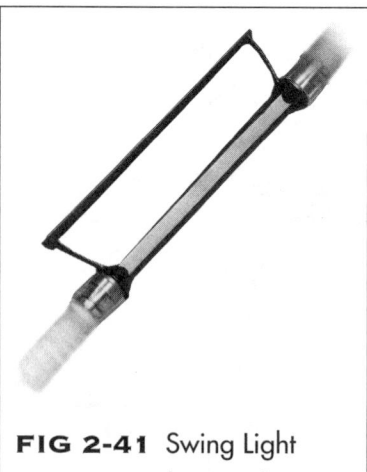

FIG 2-41 Swing Light Trainer provides visual extension of club shaft when club shaft is behind your body.

1. At address, the shaft of the club rests on an incline plane. The club head rests in front of your body.

2. During the golf swing, the club head moves behind your body. To swing the club "on-plane," you must try to maintain the same incline plane inside the training zone.

3. Inside the training zone, when the Swing Light Trainer points to the Plane Line, you are on-plane.

Remember: The forward arm pulls down the vertical plane while the club shaft travels around your body on an incline to meet the ball from the inside out.

**FIG 2-42** A. Too flat

**FIG 2-43** B. Perfect

**FIG 2-44** C. Too steep

Most players try to use their forward arm to hit a ball from the inside out during the downswing. This is not good. This action results in a probable slice into an adjacent fairway. The forward arm must sweep parallel to the Plane Line through the training zone. The forward arm must stay on the vertical plane through the Training Zone.

During the backswing, the swing light must keep in contact with Blanchard's Plane Line until the forward arm reaches about 8 o'clock on an imaginary clock face.

The shaft of the club should hinge upward by 9 o'clock on the clock face. Then, it's the swing light from the grip end of the shaft that must pick up the Plane Line and stay connected to the top of the backswing.

The club "headlight" tracks the Plane Line until the forward arm has reached 8 o'clock on an imaginary clock face.

Before the forward arm reaches 9 o'clock, the grip light must pick up the Plane Line to set the exact incline plane.

Always remember, it is the club head that strikes the ball from the inside out. This occurs naturally when the shaft of the club holds the correct incline plane on the downswing.

**FIG 2-45** 1. Club head in front

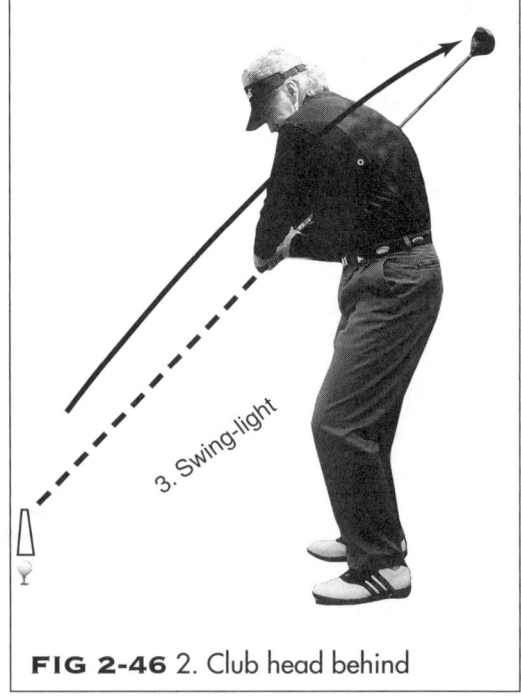

3. Swing-light

**FIG 2-46** 2. Club head behind

**FIG 2-47** 1. Swing Light
"Head-light"

*Source:* Adapted and reprinted
with permission from
Anne Fewell.

**FIG 2-48** 2. Swing Light
"Grip-light"

*Source:* Adapted and reprinted
with permission from
Anne Fewell.

Imagine the forward arm pulling up, then down the vertical plane with the trailing club head on the incline plane. The successful execution of these two moves will lead to powerful and solid contact with the ball.

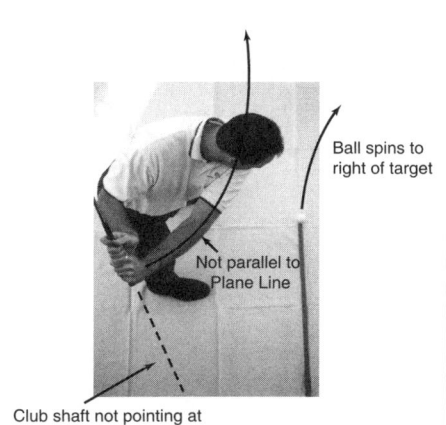

Ball spins to right of target

Not parallel to Plane Line

Club shaft not pointing at Plane Line

**FIG 2-49** When the forward arm is not parallel to the Plane Line, the arm will rotate around the body through the training zone.

*Source:* Adapted and reprinted with permission from Anne Fewell.

Shaft around

Arm up vertical plane

**FIG 2-50**

*Source:* Adapted and reprinted with permission from Anne Fewell.

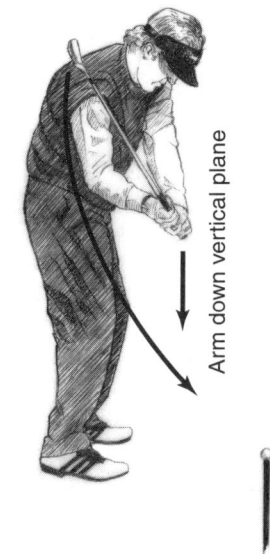

Arm down vertical plane

**FIG 2-51** As the forward arm goes up and down the vertical plane, the club shaft rotates around the body on the incline plane.

*Source:* Adapted and reprinted with permission from Anne Fewell.

# THE SWING PLANE

**Definition:**
*1. The movement of the forward arm on the vertical plane.*
*2. The movement of the club shaft through the incline plane.*

At the top of the backswing, the very same preset angle at address should invert itself as the club shaft moves up and behind your torso along the incline plane (Figures 2-52 and 2-53).

## KEY POINTS

For every club in your bag, there are two swing planes.

The idea that there is a single swing plane misleads the golfer into rotating the forward arm around and behind the torso during the backswing. This sets up the dreaded spin move on the downswing. The spin move creates an outside-to-inside swing path across the face of the ball, producing the dreaded slice.

Only the shaft of the club should get behind your torso on the backswing. The forward arm needs to stay out in front of the body and parallel to the Plane Line. The forward arm travels on the vertical plane while the shaft of your club travels around your body on the incline plane.

**FIG 2-52** The club shaft needs to stay on the same plane as it rotates around the body.

SAND WEDGE
Shaft behind body

**FIG 2-53**

54

The important point to remember is that the forward arm and the club shaft are connected and swing together but in two different planes: inside the training zone, the forward arm swings up and down on the vertical plane. The club shaft rotates around the body on an incline plane.

There are 14 different incline planes for the club shaft—one for every club in the bag.

If there were a fourth key point, it would be regarding lag. By definition, *lag* means to proceed slowly, to fall behind. The reference for golf is to the relationship of the forward arm and club shaft. Once the forward arm has climbed to 9 o'clock on an imaginary clock face, the club shaft should be at a 90 degree angle to the extended forward arm and on the correct incline plane behind the body (Figure 2-55).

The 90-degree relationship (between the forward arm and club shaft) should be maintained to the top of the backswing and

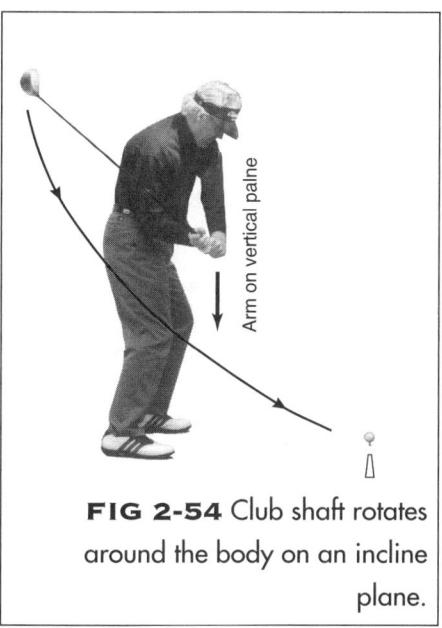

**FIG 2-54** Club shaft rotates around the body on an incline plane.

**FIG 2-55** During the downswing, maintain the 90-degree angle formed between the club shaft and the forward arm until the forward arm has dropped to 8 o'clock on an imaginary clock face.

When done correctly, this swing causes the club head to lag behind the body as the golfer pulls the heel of the club shaft down the Plane Line into the back of the ball (Figure 2-57).

continue to be maintained until the forward arm has reached 8 o'clock on the downswing.

If you are diligent with your training, you notice this will happen automatically if you keep the heel light of your Swing Light Trainer pointing at the Plane Line behind the ball until the forward arm has descended to 8 o'clock on the imaginary clock face.

Unfortunately, most amateurs release (open) that critical 90-degree angle the moment they begin their downswing. This is called *casting*. When golfers permit the early release of the club head, they lose power and immediately move the club shaft off-plane.

**FIG 2-56** The baseball "lags" behind the release of the hips and shoulders of the pitcher.
*Source:* © Aun-Juli Riddle/ShutterStock, Inc.

**FIG 2-57** The club head "lags" behind the release of the shoulders, arms, and hips of the golfer.

# MECHANISM OF INJURY

**Definition:**
*Mechanism is a system of parts that interact. Injury is a physical harm.*

---

## KEY POINTS

The three key points of the golf swing are the individual parts of an integrated system of physical movement.

When golfers violate one or more of the three key points, they put their bodies at risk of injury.

---

## KEY POINT 1: THE BACK FOOT IS PERPENDICULAR TO THE PLANE LINE

When the back foot is perpendicular to the Plane Line behind the ball, the golfer will ensure a directional release of energy directly toward the intended target.

With the back foot turned outward or open to the Plane Line, as the golfer pushes off the back foot to start the downswing, he or she pushes the energy out to the right of the intended target. This generally forces the golfer to speed up the release of the club head. Said another way, the golfer is forced to cast the club head out and over the top of the swing plane to try to close the club face at impact; otherwise, the ball pushes out to the right of the target.

An early release of the club head is a golfer's effort to reroute the energy of the body and the club head back toward the target. Remember, the club head is traveling at about 90 mph. This attempt at correction, midswing, puts unwanted stress on the muscles, tendons, ligaments, and joints of the neck and shoulders. After hundreds of swings like this, your patients will develop a repetitive strain syndrome.

In addition, the open back foot contributes to the back knee rotating outward during the backswing. The back knee should be completely stable and remain pointing straight ahead at the Plane Line during the backswing. If your patients' back knee is rotating back and forth during the backswing and the downswing, you can expect them eventually to develop knee pain from the excess and unnecessary movement.

The same rule applies for the back ankle. When the back foot is perpendicular to the Plane Line, the weight can stay on the inside of the back foot to the top of the backswing. With the back foot open, the weight rolls onto the top and sometimes toward the outside of the back foot. This excess movement is not only unstable, it can lead to repetitive strain of the back ankle.

## KEY POINT 2: THE FORWARD ARM MUST SWEEP PARALLEL TO THE PLANE LINE

When a golfer is swinging safely, the kinetic energy is released directly toward the intended target. Because the Plane Line is contiguous with the target line, it should make sense that the forward arm should sweep parallel to the Plane Line to release the energy of the body directly toward the target.

When the forward arm rotates around the body relative to the Plane Line, it's anyone's guess exactly when the energy will be released. An early release of energy pushes to ball out to the right of target. A late release of energy pulls the ball over to the left of target.

More important, golfers will have an intuitive sense they are out of position. By reflex, any time the forward arm is not sweeping parallel to the Plane Line, the body will attempt to compensate by attempting to delay an early release or to speed up a late release. Don't forget, the club head is already in motion around the body, traveling at some 90 mph. These sudden changes of direction directly affect muscles, tendons, ligaments, and joints of the body. After hundreds of swings, the end result can be a repetitive strain syndrome.

Keep an eye out for the golfer who bends that forward arm during the backswing. This golfer bends the arm because he or she has reached the elastic barrier of the connective tissue that surrounds the scapulothoracic and glenohumeral joints. During the downswing, centrifugal force of the club head snaps that bent forward elbow into extension. After hundreds of swings, you have a repetitive strain syndrome secondary to repeated hyperextension of the forward elbow.

**FIG 2-58** When the back foot is perpendicular to the Plane Line, the back knee and back ankle are stable and less prone to injury.

## KEY POINT 3: THE CLUB SHAFT MUST ALWAYS POINT AT THE PLANE LINE

When the club shaft points at the Plane Line behind the ball, the golfer is swinging on-plane. This is good . . . very good.

The most common injury associated with allowing the club shaft to move around your body off-plane is wrist strain.

Again, we are back to following the directional release of energy. If the club shaft is too flat on the downswing (this means the heel light from the Swing Light Trainer is shining away from your body, out and over the top of the Plane Line) the directional release of energy is heading to the right of your target. To compensate, golfers flip their wrists at the last minute in an effort to reroute the flight of the ball back to the left. This flip motion snaps the back wrist (that is the right wrist for the right-handed golfer) into a state of hyperflexion.

When the club shaft gets too steep on the downswing, the heel light of the Swing Light Trainer shines toward the body, inside the Plane Line, or closer to the feet. When this happens, the golfer runs the risk of smacking the club head into the ground behind the ball.

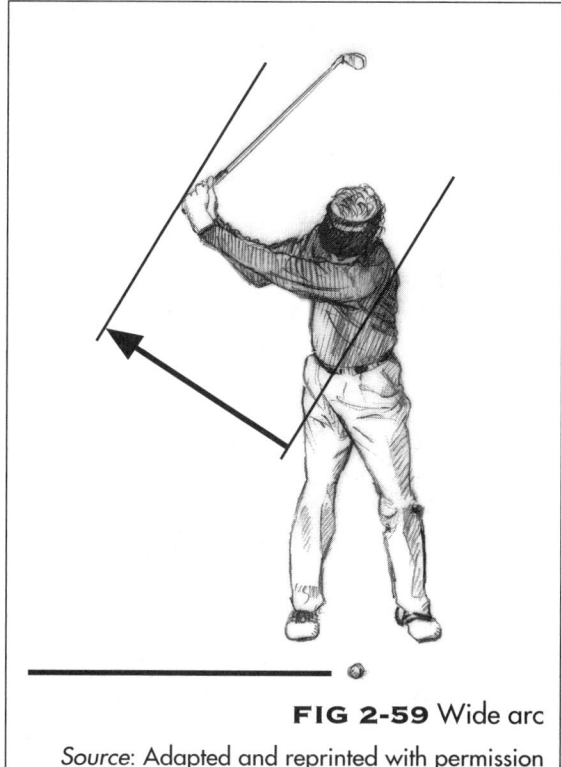

**FIG 2-59** Wide arc

*Source:* Adapted and reprinted with permission from Anne Fewell.

**FIG 2-60** Collapse of arc

*Source:* Adapted and reprinted with permission from Anne Fewell.

This is called "hitting it fat." Hitting "fat" creates a rapid deceleration of the club shaft relative to the hands and wrists and forces the forward wrist into hyperflexion. Repeated swings with a club shaft that is too flat relative to the Plane Line results in a repetitive strain syndrome of the forward wrist.

## SWING ANALYSIS

Have your patient take a club and make a few practice swings (Figures 2-62, 2-63 & 2-64). You could provide a synthetic grass

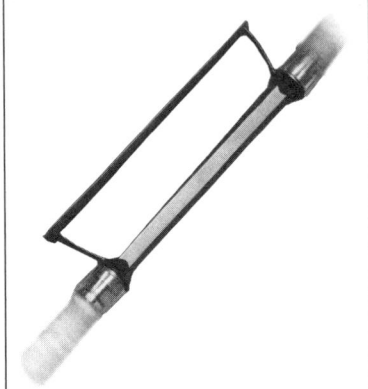

**FIG 2-61** Swing Light Trainer provides visual extension of club shaft when club shaft is behind your body.

**FIG 2-62** A. Too flat

**FIG 2-63** B. Perfect

**FIG 2-64** C. Too steep.

mat and some foam golf balls if you want, but this is not necessary. If your treatment rooms have 8-foot ceilings, take an old golf club to a pro shop and have them cut it down and put on a new grip. The length of club, however, is irrelevant to your analysis.

You must look for and identify three different components of the patient's golf swing.

There are too many moving parts of body and club to try to do otherwise.

> The Swing Light Trainer is an effective diagnostic tool to determine if your swing plane is too flat, just right, or too steep.

## THREE KEY POINTS

Is the position of the back foot perpendicular to the Plane Line?

Does the extended forward arm trace parallel to the Plane Line during the backswing and the downswing?

Does the club shaft point to the Plane Line throughout the swing?

"That's it?" Yes, that's it. "Hey, but what about . . . ?" Every single "what about" takes care of itself if your patients train to perfect execution these three key points.

1. Be sure your back foot is perpendicular to the Plane Line.

The position of your right foot dictates the direction of the flight path of the ball. In addition, when you square up your right foot to the Plane Line, your hips and shoulders coil quickly during the backswing. A strong, tight coil *with your weight inside the back foot* is a necessary step to generate kinetic energy for a powerful downswing.

Before you swing the club, be sure that your hips and shoulders are parallel to the Plane Line.

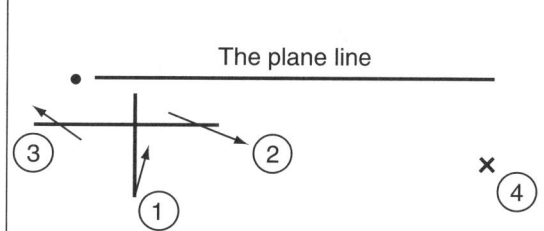

**FIG 2-65** 1. The back foot is positioned open to the right.

2. Backswing: forward arm tracks inside the Plane Line.

3. Downswing: forward arm pushes toward the Plane Line.

4. At the top of the swing, the heel light from the Swing Light Trainer shines inside the plane line (see the X). This means the incline plane for the shaft of the club is too steep.

When you watch your patient swing the club, you must train yourself to look at one component at a time.

This ensures proper alignment of your body before you swing the club. Well over 80 percent of players are set up closed to the Plane Line at address. Before they ever swing the club, the ball is already going to the right of their intended target. When swinging the club:

2. Keep your forward arm parallel to the Plane Line.

3. Always point the club shaft at the Plane Line.

With the forward arm parallel to the Plane Line, you can coil and then release your stored kinetic energy directly toward your intended target. When the club shaft points at the Plane Line, you will be on-plane. In addition, you are assured of striking the golf ball from the inside out.

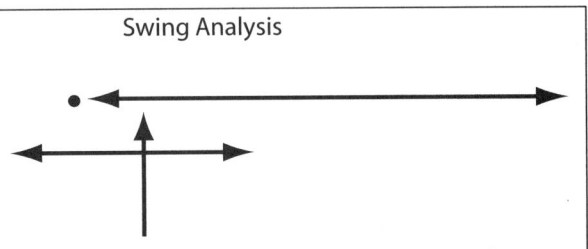

Swing Analysis

**FIG 2-66** Train your patients to stand and swing like this. Now, all the kinetic energy will coil into the correct position and release safely and effectively toward the intended target. This pattern will help your patients swing pain-free and play better golf.
*Source*: Adapted and reprinted with permission from Anne Fewell.

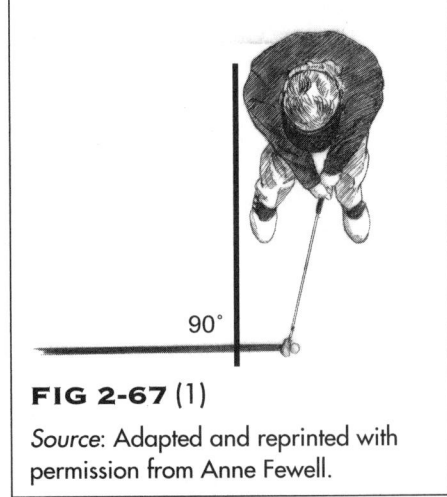

90°

**FIG 2-67** (1)
*Source*: Adapted and reprinted with permission from Anne Fewell.

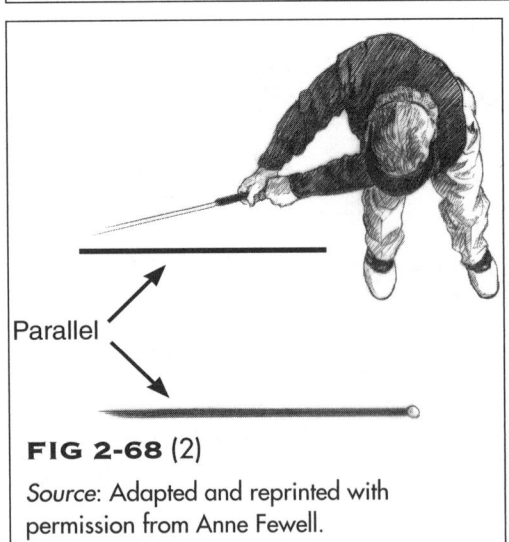

Parallel

**FIG 2-68** (2)
*Source*: Adapted and reprinted with permission from Anne Fewell.

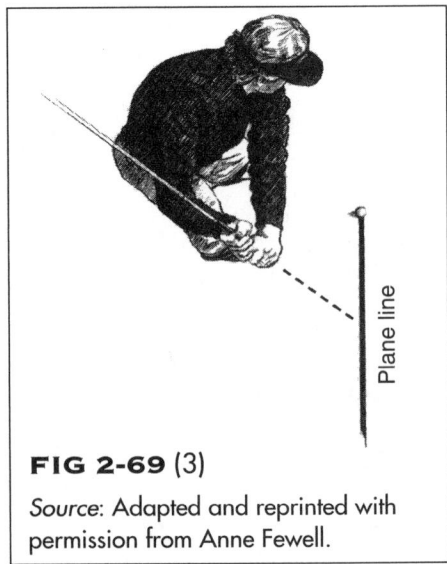

Plane line

**FIG 2-69** (3)
*Source*: Adapted and reprinted with permission from Anne Fewell.

# THE GEOMETRY OF A SLICE

It has been said that approximately 80 percent of every golf shot hit off-line is in the form of a slice to the right of the intended target.

To begin with, make sure your key points are in place during the downswing.

A.

**KEY POINT 1:** Back foot is perpendicular to the Plane Line. This permits the energy from the leg drive to release toward the target.

> *A slice occurs whenever the club face strikes the ball open relative to the Plane Line behind the ball (Figure 2-70).*

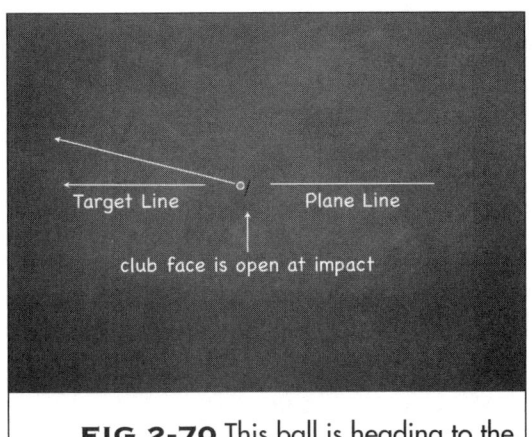

**FIG 2-70** This ball is heading to the right of target.
*Source:* Adapted and reprinted with permission from Anne Fewell.

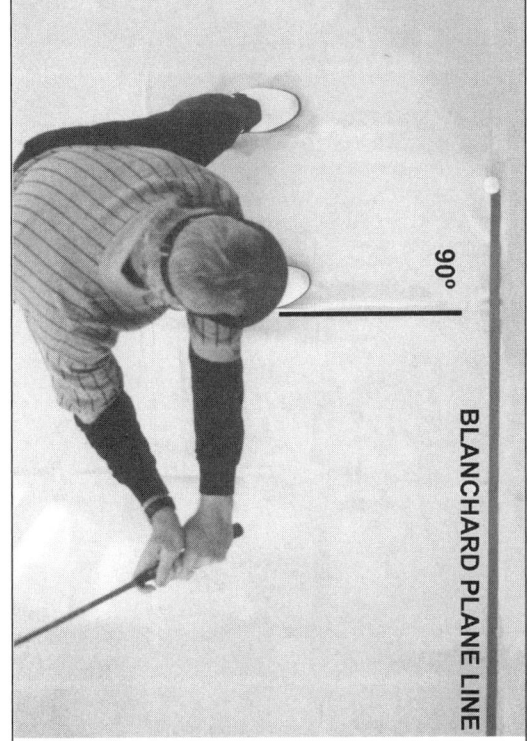

**FIG 2-71** Correct position
*Source:* Adapted and reprinted with permission from Anne Fewell.

If your patients have a chronic slice, make sure they "close the door." After impact, they must get the heel light of the club shaft around and pointing at the target line in front of the ball.

The key to correction of a slice is to "mirror" the *before* impact downswing positions immediately *after* impact with the ball.

**KEY POINT 2:** The forward arm is parallel to the Plane Line. This permits the energy of the arm swing to release toward the target.

**KEY POINT 3:** The club shaft is pointing at the Plane Line. This permits the club shaft to remain on the correct incline plane.

## B.

**KEY POINT 1:** Back foot is ***not*** perpendicular to the Plane Line. This permits the energy from the leg drive to release right of the target.

**KEY POINT 2:** The forward arm is ***not*** parallel to the Plane Line. This permits the energy of the arm swing to rotate/spray anywhere relative to the target.

**KEY POINT 3:** The club shaft is ***not*** pointing at the Plane Line. This permits the club shaft to drift off the necessary incline plane.

After impact, it is critical that you mirror your downswing positions.

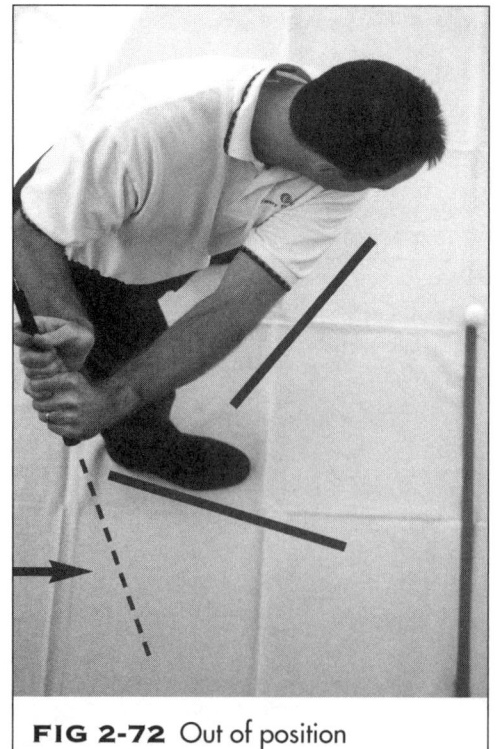

**FIG 2-72** Out of position

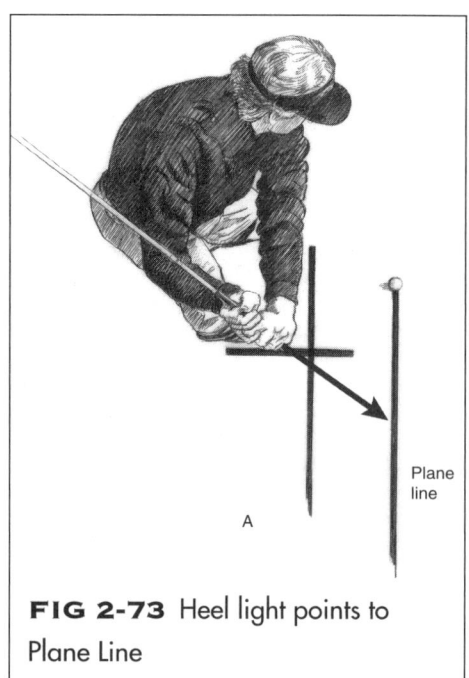

**FIG 2-73** Heel light points to Plane Line

## A.

The forward arm is parallel to the Plane Line, from about 9 o'clock to 6 o'clock position.

When you use a Swing Light Trainer, the heel light of the club shaft points to Plane Line before impact.

## B.

After impact, the back arm is parallel to the target line, from about 6 o'clock to the 3 o'clock position.

When you use a Swing Light Trainer, the heel light of the club shaft points to target line after impact.

## A NEW PERSPECTIVE

A. Successful execution of key points 1, 2, and 3 relative to the Plane Line *before* impact (Figure 2-75).

B. Successful mirror image relative to the target line *after* impact (Figure 2-76).

The majority of patients find this difficult to do because it requires exceptional flexibility in lateral bending, rotation, and extension in the thoracolumbar spine to get into the after-impact position.

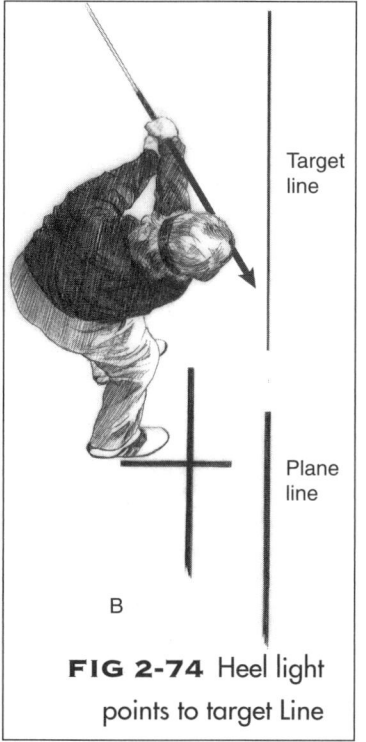

Target line

Plane line

B

**FIG 2-74** Heel light points to target Line

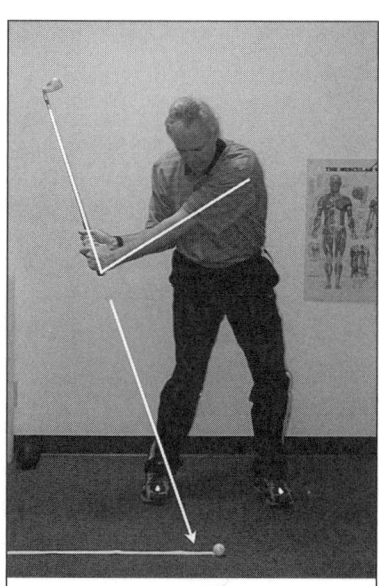

**FIG 2-75** A Before impact, Plane Line

**FIG 2-76** B After impact, Target Line

# Chapter 3

## BARRIERS TO PERFORMANCE

### Well Patient Evaluations and Management Protocols

# INTRODUCTION

Let's take a look at the most common sites of injury in amateur and professional golfers.

## AMATEURS

**LOW BACK** The average amateur golfer has the back foot in an open (turned-out) position relative to the Plane Line behind the ball. The open back foot permits over-rotation of the lumbosacral spine. During the golf swing, in addition to rotation, the lumbar spine is subjected to forces of lateral bending, anteroposterior shearing, and caudocranial compression. The lumbar spine of an amateur, however, undergoes significantly more loading than that of a professional because of lack of golf-specific conditioning and poor balance.

*Amateurs: Almost 70 percent of all golf injuries are either low back or elbow injuries.*

After 18 holes of golf or hundreds of swings on the driving range, is it any wonder an amateur's low back hurts? Remember, the reason most amateurs turn the back foot out at the address position is because the person already has poor flexibility and physical restriction in the lumbopelvic region.

**ELBOWS** During the backswing, when the patient reaches the elastic barrier of the connective tissue surrounding the scapulothoracic and glenohumeral joints of the forward arm, the patient bends at the elbow to keep moving the club head farther behind the body. During the downswing, the bent elbow snaps into hyperextension at impact with the ball. Ouch!

## PROFESSIONALS

**HANDS/WRISTS** Golf professionals injure their hands and wrists because they are required to play the ball "down." In other words, if their ball is in the deep, wet rough or in a bush or adjacent to a tree root, they cannot move the ball without taking a penalty stroke. So, they hit golf balls from places most amateurs will not hit from. The injury occurs as a result of the rapid deceleration of the club head relative to the connective tissues in the hands and wrist as the club head meets the resistance of the deep grass, the bush, or the tree root.

*Professionals: About 35 percent of golf injuries are hand/wrist injuries, and about 20 percent are low back injuries.*

## FRACTURES

Fractures in golfers are uncommon, but under the conditions noted previously, fracture of the hook of the hamate ("golfer's wrist") is the most common fracture in golf and often goes undiagnosed for several weeks to months.

The most likely cause of a fracture is a "fat" shot (even when hitting from the fairway, the club head buries into the ground behind the ball). Anytime the club head strikes a relatively heavy or immovable object, such as a rock or tree root, the golfer is at risk of fracture because the butt of the club is forced against the hypothenar region of the leading hand, and because the club head is traveling at 90 mph or faster, breaking the hook of the hamate is a very real possibility.

Be sure to ask any patient with complaints of hand and/or wrist pain whether he or she has hit any shots that might have applied an undo amount of force to the structure in question.

## "IT REALLY HURTS, BUT I REALLY WANT TO PLAY"

Short of when lightning strikes on the golf course, golfers will continue play, injured or not. At times, you will have to get injured golfers to rest for a few days (whether they like it or not). The following "Is it OK to play?" guidelines are provided to help you make such a recommendation.

### Stop Playing!

Whenever the injured body part continues to hurt at rest—it's over, at least for a while. Put the clubs away.

### Minimal Activity

Chipping around the green and putting is recommended once the injured body part no longer hurts at rest. Remind your patients that chipping and putting account for more than 50 percent of their score, and they could probably use the practice.

### Resume Play

If your patient can take full *practice* swings without pain, have him or her begin hitting foam or whiffle balls about the yard. If the patient can tolerate this, he or she can start hitting shots in the 100- to 150-yard range for a few days. The agreement must be to stop if the pain returns.

## ICE AND STRETCH

- Ice the entire muscle until it's numb. Trainers often fill small paper cups with water and freeze them. Then, peel down the paper to expose the ice and rub it on the muscle until the muscle is numb.

- Then, have the patient gently move the numb muscle until he or she begins to feel some pain. Ask the patient to hold it there, at the threshold of pain, for about 20 seconds. After about 20 seconds, have the patient try a 5-second isometric contraction of the numb muscle.

- Have the patient immediately begin to move the muscle again until he or she feels the threshold of pain again. Hold it there for 10 seconds this time, and then have the patient try another 5-second isometric contraction.

- Let the patient rest for about 30 seconds, and then repeat the process.

## SORE MUSCLES

When the body is not conditioned for golf, hitting 100 balls at the range can produce sore muscles. In addition, slow play over a

5-hour round of golf permits the body to get stiff and tight between shots. Encourage your patients to walk the golf course instead of riding in a golf cart. Walking elevates the body's core temperature, which helps decrease stiffness. General soreness is best treated with light stretching and a good heat-producing liniment. (See Play On Pain Relief Lotion, described in the section titled "Anatomy of a Golf Injury" in Chapter 1.)

## MUSCLE SPASM

Muscle spasms generally appear hours, or even days, after a session at the driving range or a round of golf. To the patient, the onset of the spasm is acute; however, the root cause is often insidious. *Note:* If your patient has suffered a muscle tear, the pain would be immediate, and the person would know there has been an injury and can tell you exactly when it happened. Muscle spasms, from overuse, begin with little more than localized pain and stiffness and might include a slight loss in range of motion. Left unresolved, the onset of spasm is common. Your patients can generally work through mild to moderate muscle spasm using the Ice and Stretch protocols outlined earlier.

## MUSCLE STRAINS

Failure to warm-up before hitting balls on the range or before a round of golf predisposes your patients to muscle strains. A muscle will be strained whenever the fibers of the muscle are stretched beyond their elastic capacity.

A strained muscle is characterized by specific, localized pain and swelling. Once strained, injured muscle fibers need rest and ice. The body can tolerate ice for 20 minutes every hour if necessary. Ice is most effective for the first 1 to 3 days after injury or until the pain and swelling subside. Depending on severity, it might take as long as 6 weeks to recover from a muscle strain.

**FIG 3-1** South American holly shrub.

*Source:* © Giovanni Civardi from *Drawing Human Anatomy*. Reprinted courtesy of Sterling Publishing Co., Inc., New York, NY.

## ILEX

ILEX is an herbal extract from a South American holly shrub. ILEX is used around the world in various health and wellness formulations (Figure 3-1). The ideal preparation does

ILEX can aid in the treatment of acute injuries and can be used in any supervised situation that requires the use of ice or breathable wraps.

Shoulders that are generally level to the horizon must tilt at the address position.

not include wax, oil, aloe, or petroleum in the product because each impedes the effectiveness of the herb. The herb is a fast-acting, penetrating, long-lasting pain reliever.

# POSTURAL COMPROMISE

Getting into position to hit golf shots requires postural compromise.

The right hand is below the left hand on the club shaft. This pulls the right shoulder lower than the left, creating shoulder tilt at the address position.

## FORWARD SPINE ANGLE

Further evidence of postural compromise is found at the address position. The upright spine must shift

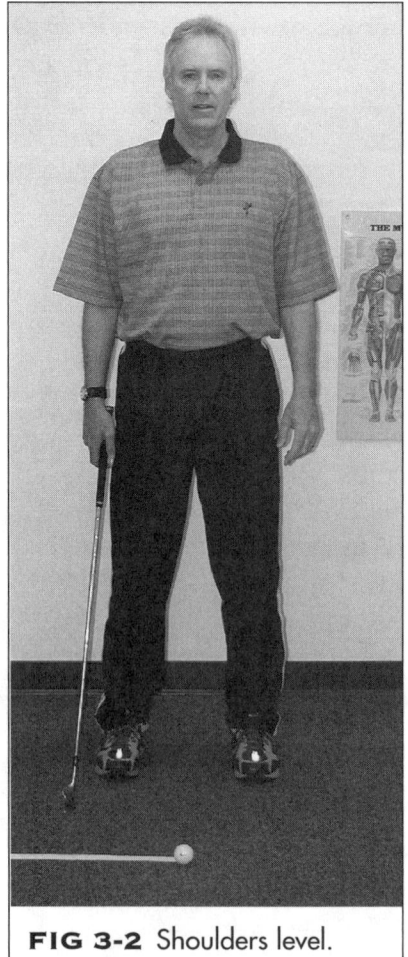

**FIG 3-2** Shoulders level.

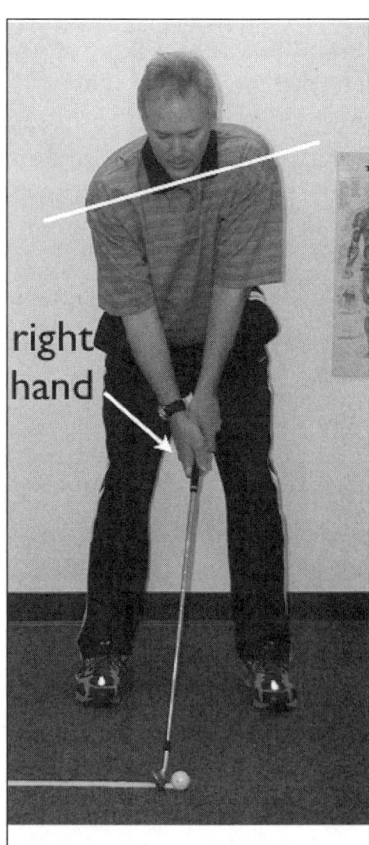

right hand

**FIG 3-3** Shoulders tilt at address position.

into a forward tilt from the waist when swinging the club around the body.

Failure to maintain the forward spine angle results in inconsistent and poor-quality golf shots (Figures 3-4 and 3-5).

## GOLF-SPECIFIC CONDITIONING

Unfortunately, most golf patients have not conditioned their body for this demanding sequence of movements. Especially after impact, if the elastic barriers in the lumbopelvic region have not been lengthened, patients will begin to stand up moments before impact to avoid pain.

The forward tilt of the spine relative to the vertical plane is called the *spine angle*. To hit quality golf shots it is necessary to maintain the spine angle during the backswing and during the downswing.

To swing the club around the body *and* maintain the necessary spine angle requires the spine to bend laterally, rotate, and extend all at the same time (Figures 3-6 and 3-7).

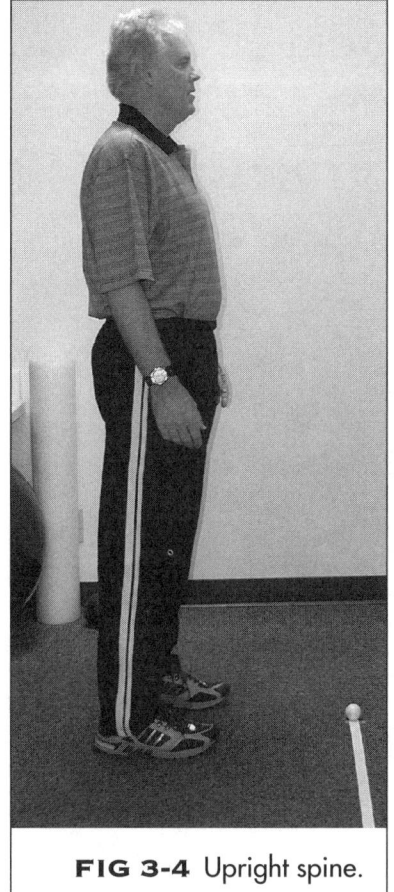

**FIG 3-4** Upright spine.

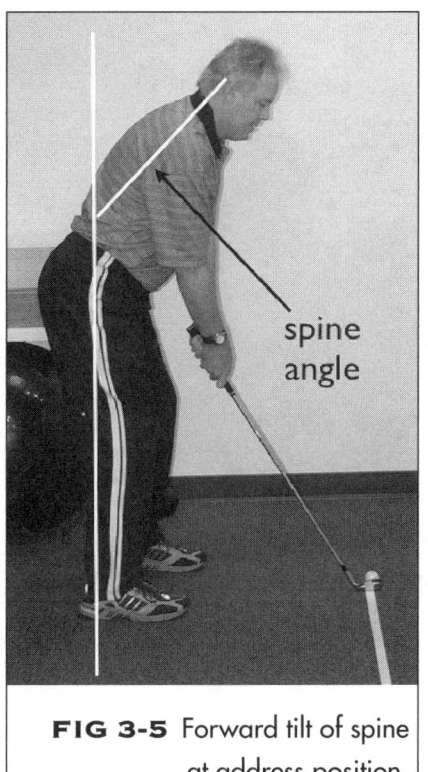

spine angle

**FIG 3-5** Forward tilt of spine at address position.

When a patient stands up just before impact with the ball, it is referred to as "coming out of a shot."

As a rule, when a patient stands up just before impact, the shot will slice off to the right of target.

**CORRECTION PROTOCOL:** If your patients cannot bend lateral, rotate, and extend their spine without painful restriction, see related spinal range of motion correction protocols in Chapter 1: Physical Examination.

## BALANCE

A 250-yard drive is no longer good enough!

Golfers want to hit big, powerful drives. With the new generation of strong, athletic golfers in the PGA and the LPGA, 250 yards off the tee is no longer good enough. On any given day, your patients are also on the golf course or at the driving range trying to drive the ball 300 yards.

Paul Chek, in his excellent book *The Biomechanics of Golf*, writes: "Amateur golfers achieve approximately 90 percent of their peak muscular effort when driving a golf ball. This is the same intensity as picking up a weight that can only be lifted four times before

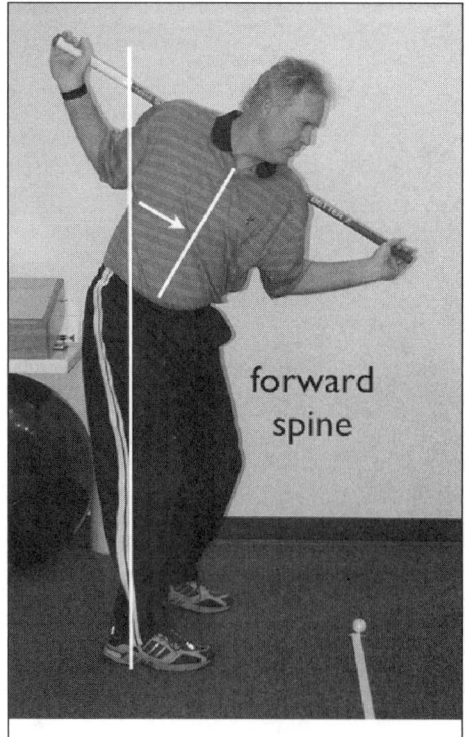

**FIG 3-6** During backswing, maintain forward spine angle.

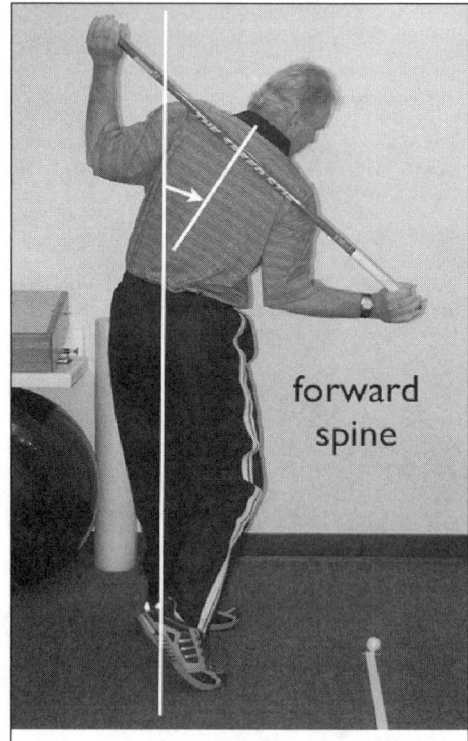

**FIG 3-7** During downswing, maintain forward spine angle.

total fatigue." And that's just for one swing! Golfers take hundreds of swings when they play or practice golf.

Golfers have a high incidence of injury The injuries are of the repetitive strain variety. Your patients swing their golf clubs at speeds of 90 to 115 miles per hour. When they swing off balance, there is undue and often violent strain on the muscles, tendons, ligaments, and joints of the body.

So, how do you, as the treating physician, identify issues of balance in the golf swing? It begins with understanding the *Plane Line*, an imaginary line on the ground that starts just behind the golf ball and runs 10 feet behind the ball. There are four key points to identify and diagnose when your patients move their body and swing the club relative to the Plane Line (Figures 3-8 through 3-11).

To swing on balance means that as you must generate kinetic energy during your backswing and release the stored energy down

It has been reported that nearly 50 percent of golfers will eventually suffer a golf-related injury.

An off-balance golf swing implies that your patient has lost the necessary positional relationship and alignment to the Plane Line.

At address, the right foot should be perpendicular to the plane line.

**FIG 3-8**

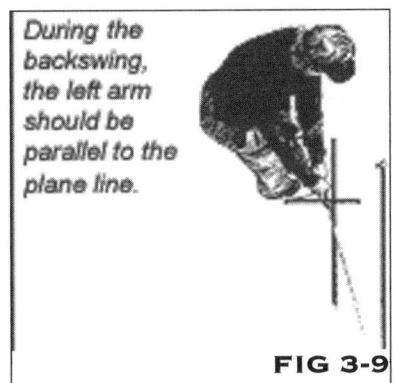

During the backswing, the left arm should be parallel to the plane line.

**FIG 3-9**

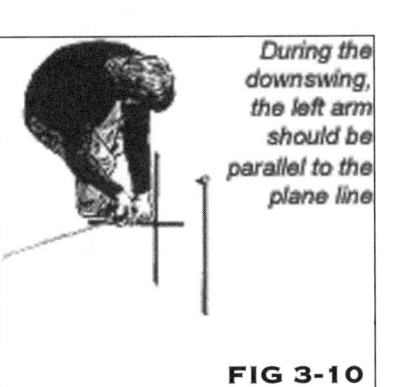

During the downswing, the left arm should be parallel to the plane line

**FIG 3-10**

To set the correct incline plane, the shaft of the club must always point to the plane line.

**FIG 3-11**

the Plane Line through the ball and directly toward your intended target.

During the downswing of an off-balance swing, the stored kinetic energy is not released toward the intended target. Your patient's body reacts by reflex once it realizes it's out of position. Watch compassionately as your patients flip their hands and overreach, push and pull with the arms, and then lunge at the ball. They will do whatever it takes to "get that darn ball to behave!" However, the muscles, tendons, ligaments, and joints eventually become strained from the abuse (Figures 3-12 and 3-13).

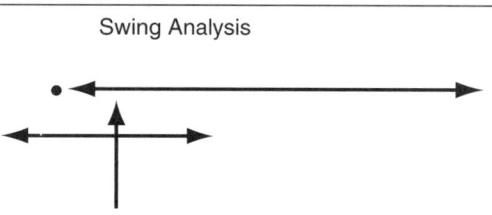

Swing Analysis

**FIG 3-12** Perfect tracing of a balanced golf swing. The back foot is positioned at 90 degrees to the Plane Line. The forward arm tracks parallel to the Plane Line. The club shaft is always *pointing to the* Plane Line.

*Source*: Adapted and reprinted with permission from Anne Fewell.

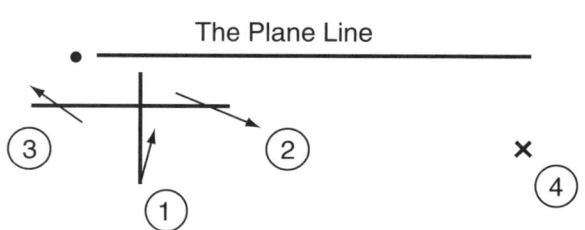

The Plane Line

**FIG 3-13** Off-balance tracing of the golf swing.

1. The back foot is positioned open to the right.

2. Backswing: forward arm tracks inside the Plane Line.

3. Downswing: forward arm pushes toward the Plane Line.

4. At the top of the swing, the heel light from the Swing Light Trainer shines inside the plane line (see the X). This means the incline plane for the *shaft of the club* is too steep.

*Source*: Adapted and reprinted with permission from Anne Fewell.

**CORRECTION PROTOCOL** Your patient must undergo repetitive, yet objective, neuromuscular reeducation. The goal is to retrain the body to release the stored kinetic energy of the backswing safely and efficiently down the Plane Line, through the ball, and toward the intended target. This will minimize and, in many cases, eliminate the unwanted strain on the muscles, tendons, ligaments, and joints of the injured golfer.

Have your patient tape the Plane Line on the floor, and rehearse his or her backswing and downswing with the left arm parallel to the Plane Line between 6 o'clock and 9 o'clock on an imaginary clock face. (They do not need to hit balls to rehearse. In fact, it's better that they do not hit balls while reeducating the neuromuscular pathways and motor memory.) They will need a device known as a Swing Light Trainer to be able to practice tracing the incline plane of the club shaft relative to the Plane Line (Figure 3-14). This should be repeated 30 to 50 times daily for 3 weeks. After 3 weeks of perfect rehearsal, your patient will have a new golf swing.

> When you can see the root cause of the repetitive strain syndrome, it is your responsibility to identify the problem to the patient and offer a corrective solution.

## FOOTWORK

Good footwork is essential to playing better golf—but what is good footwork, and how can it be evaluated and, when necessary, treated?

Good footwork begins with feet that are fit, functional, and mechanically sound. The feet are designed to point straight ahead; each foot is designed to be directly beneath the knee.

When the feet are out of positional alignment, unwanted torque occurs in the gait muscles, beginning in the hips. In this scenario, the body must introduce compensating motion with every step (Figure 3-15).

Supination and pronation influence the stability of the pivot during the backswing. Seventy-five percent of the body weight transfers directly into the quadriceps of the back leg at the top of the backswing. The back foot must be able to support this weight transfer, and then push the weight to the front foot during the downswing.

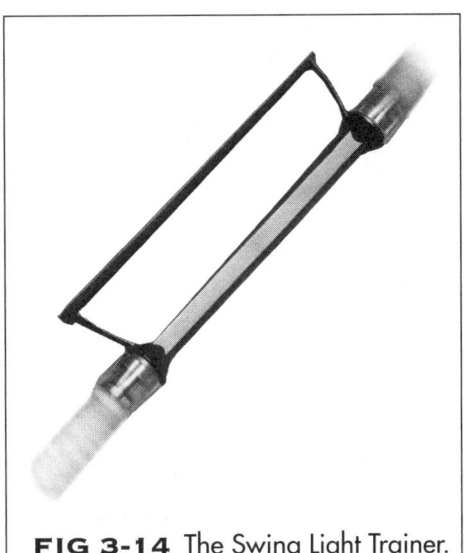

**FIG 3-14** The Swing Light Trainer.

Additionally, a golfer walks an average of 3 to 5 miles over a period of 4 to 5 hours while completing 18 holes of golf. With conditions of supination, pronation, bunions, or hammertoes beneath the ankle, the golfer risks chronic instability when attempting to transfer the weight during the golf swing or walk without compromise (Figures 3-16 and 3-17).

## RECOMMENDATIONS

Many diversified chiropractic adjusting techniques work well to reduce physical restrictions in the ankles and feet. If your technique needs polishing, you can choose from several excellent postgraduate seminars on manipulation. Contact individual state chiropractic associations for details.

Orthopedic appliances (supportive shoe inserts) are recommended for any golf patient diagnosed with chronic mechanical instability of the knee, ankle, or foot. (*Note:* Many professional golfers now use or-

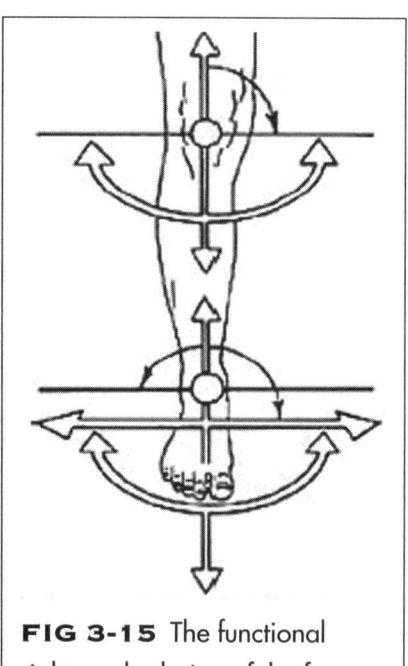

**FIG 3-15** The functional right-angle design of the foot and ankle.

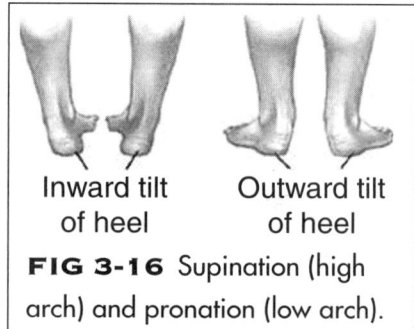

Inward tilt of heel    Outward tilt of heel

**FIG 3-16** Supination (high arch) and pronation (low arch).

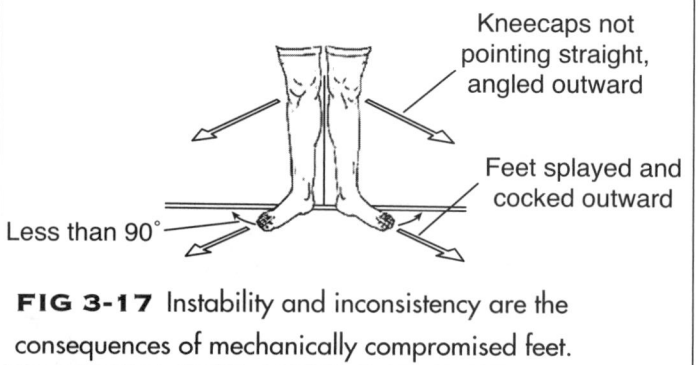

Kneecaps not pointing straight, angled outward

Feet splayed and cocked outward

Less than 90°

**FIG 3-17** Instability and inconsistency are the consequences of mechanically compromised feet.

*Source:* All figures on this page, reprinted with permission from *The Escogue Method of Health Through Motion.* © 1992, courtesy of HarperCollins Publishers.

thotics. One study reports that regular use of orthotics increases the player's club head velocity an average of 3.5 mph, which translates into an additional 10–15 yards of distance off the tee! (Figure 3-18)

## STABILITY, POWER, AND BALANCE

When hitting golf balls, the biomechanical integrity of the feet plays a key role in stability, power, and balance. Stability is needed at the top of the backswing. Power is generated when the golfer pushes off the back foot toward the target. Balance is necessary after impact when the body weight transfers to the front foot. Let's review the key structures of the foot that have specific effects in playing golf.

**THE TRANSVERSE ARCH** To maintain the athletic position (knees slightly bent, hips in extension, and body weight pressed slightly forward across the heads of the metatarsals), the transverse arch of the feet must be intact. When healthy, the transverse arch

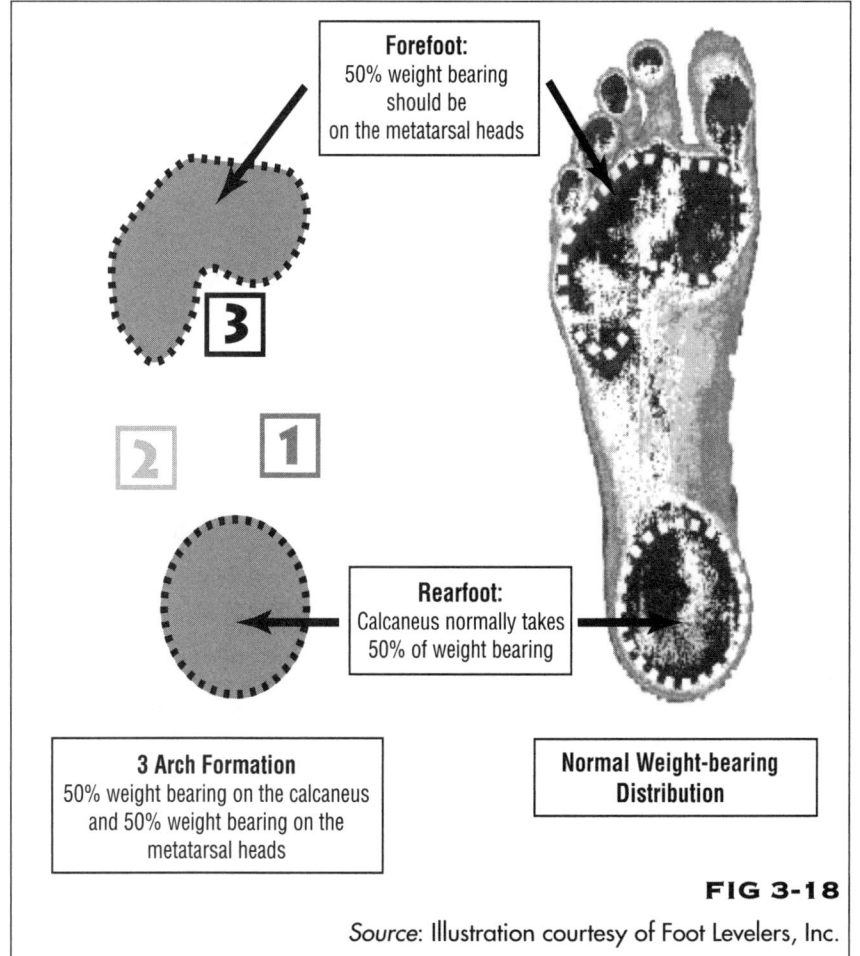

**Forefoot:**
50% weight bearing should be on the metatarsal heads

**Rearfoot:**
Calcaneus normally takes 50% of weight bearing

**3 Arch Formation**
50% weight bearing on the calcaneus and 50% weight bearing on the metatarsal heads

**Normal Weight-bearing Distribution**

**FIG 3-18**

*Source:* Illustration courtesy of Foot Levelers, Inc.

permits the toes to spread slightly and stabilize the feet during the weight transfer from backswing to downswing (Figure 3-19).

**THE LONGITUDINAL ARCH** This medial arch has two functions. It helps stabilize the knees over the ankles. During the backswing, the back knee must hold it's positional alignment over the back ankle.

In addition, the arch is necessary for the back foot to receive the transfer of body weight at the top of the backswing. The longitudinal arch must be capable of receiving 75 percent of the body weight at the top of the backswing.

**THE LATERAL ARCH** The integrity of the lateral arch permits the body weight to stay over the inside of the back foot (the medial longitudinal arch) during the backswing. If the lateral arch is unsupportive, the body weight is prone to roll over the top, and then

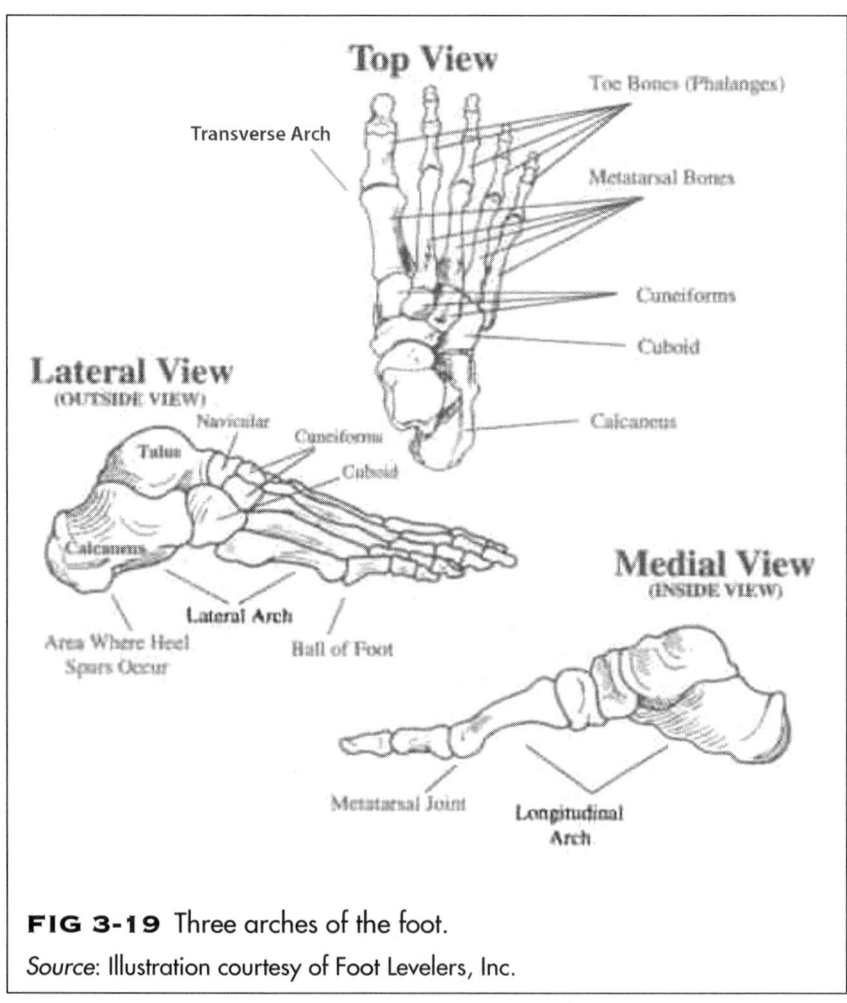

**FIG 3-19** Three arches of the foot.

*Source:* Illustration courtesy of Foot Levelers, Inc.

to the outside of the back foot. When the body weight rolls across the back foot, it creates instability and diffuses power. Over time, this instability puts the ankle, the knee, and the hip of the back leg at risk of injury.

**PHYSICAL EXAMINATION** There are three key points of physical examination: the address position, the set point at the top of the backswing, and the finish. While barefoot, have you patients get into the following three positions:

**THE ADDRESS POSITION** At the address position, the functional medial longitudinal arch should measure approximately 30 degrees, bilateral. Less than 20 degrees is inadequate for golf. When the back foot is in the proper placement the longitudinal arch should be perpendicular to the Plane Line behind the ball (Figure 3-20).

**THE SET POINT** Have your patient swing to top of the backswing, and then stop. At the top of the backswing, approximately 75 percent the body weight should load over the inside of the back foot (the medial longitudinal arch). When the body weight is

**FIG 3-20** Stability. During the backswing, the back knee must hold its positional alignment over the back knee.

**FIG 3-21** Instability. The body weight has rolled to the outside of the back foot. The back knee has rotated outward.

81

A stable platform is made of three components: proper alignment of the right knee over the right ankle, functional gluteal muscles, and strong quadriceps (Figure 3-22).

permitted to drift across the top of the foot toward the lateral arch, the kinetic energy of the coil diffuses.

**THE FINISH** Have your patient complete the swing and hold the finish position. After impact, 100 percent of the body weight should be evenly distributed through the front foot. The weight should not roll over the top or to the outside of the front foot at the finish position.

**CORRECTION PROTOCOL** For failed findings, I recommend regular adjustments to the bones of the foot. In addition, with any failed arch, I highly recommend your patient be fit with a golf-specific orthotic (Figure 3-21).

**FIG 3-22** A stable platform permits the golfer to coil body weight inside the back foot. This is an essential component to hit golf shots "on-line" to the intended target.

## STABILIZING THE PIVOT

During the backswing, your patient's shoulders and hips must pivot about the right leg. For your patient to become stable during the pivot is another matter. The next time you are at the driving range, focus your attention on the right foot of a right-handed golfer as the person swings the club. What you will see is what I call "wobble-foot." As the body weight transfers to the right side, it does not coil against a stable platform, causing wobble (Figure 3-22).

## THE GLUTEAL MUSCLES

If you imagine the kneecaps as headlights, they will aim straight ahead as long as the gluteus maximus is functional. When the gluteus maximus is weak, the femur heads will rotate internally, and your headlights will cross (Figures 3-23 and 3-24).

**CLINICAL EXAMINATION** Have your patient stand with feet pointing straight ahead and fingertips placed over the gluteus maximus. Ask the patient to make a hard gluteal contraction. Do not let the patient recruit the quadriceps or abdominal muscles. Many patients are unable to isolate and contract their gluteal muscles. When done correctly, the patient will experience external rotation of the kneecaps.

**CORRECTION PROTOCOL** Have the patient perform 30 to 50 standing gluteal contractions each day. Have the patient continue until the gluteal muscles are capable of making a strong contraction that rotates the femur heads externally (Figure 3-25).

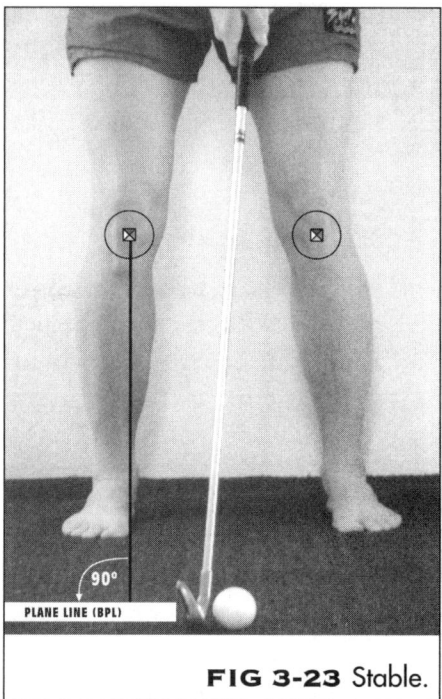

**FIG 3-23** Stable.

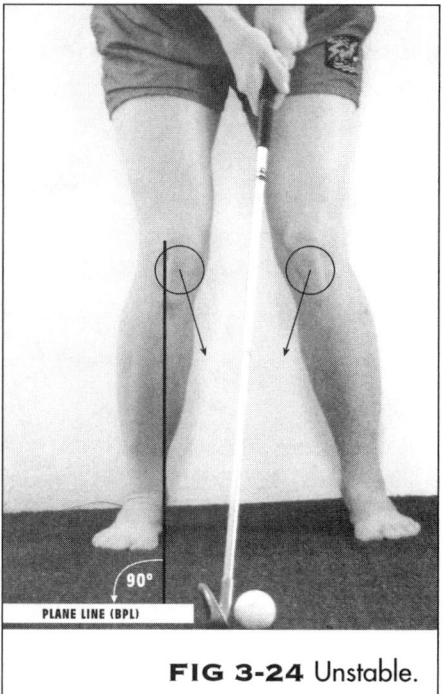

**FIG 3-24** Unstable.

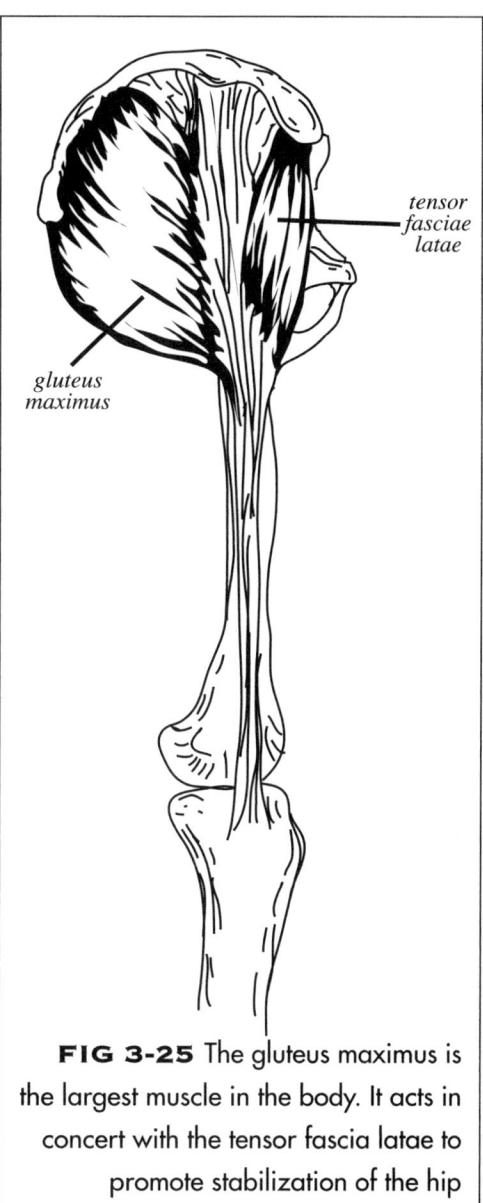

**FIG 3-25** The gluteus maximus is the largest muscle in the body. It acts in concert with the tensor fascia latae to promote stabilization of the hip and knee joints.

*Source*: © Giovanni Civardi from *Drawing Human Anatomy*. Reprinted courtesy of Sterling Publishing Co., Inc., New York, NY.

## THE QUADRICEPS

When your patient reaches the top of the back swing, 75 percent of the body weight has been transferred onto the right quadriceps. If the quadriceps are weak, your patient will not be able to hold the coil. Golfers who walk (rather than using a golf cart) cover about 4 miles on a typical golf course and "backswing" their body weight onto the quadriceps about 100 times, which includes practice swings. Therefore, the quads need to be firm, fit, and functional to play good golf.

**CLINICAL EXAMINATION** Have your patient place the heels against the baseboard of a wall. Next, move both feet outward to the length of the right foot. Make sure the feet are positioned hip-width apart. Have the patient slowly bend the knees until the patient can't see the tips of the toes. Once in position, have the patient lift the toes off the floor (Figure 3-26). Patients with functional quadriceps should be able to hold this position for 3 minutes.

**CORRECTION PROTOCOL** Have your patient perform the air bench exercise (see Part 1) daily until he or she can hold the position for 3 to 5 minutes with no discomfort.

## CONCLUSION

When your patients are at the top of their backswing, they must be able to coil against stable platforms. This requires proper position of the right knee over the right ankle as the body weight coils inside the right foot. The knee position is linked to the function of the gluteus maximus. To hold the coil with stability, the quadriceps must be functional.

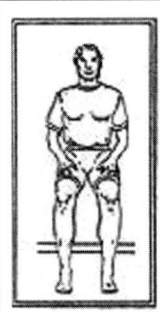 LIFT AND HOLD TOES UP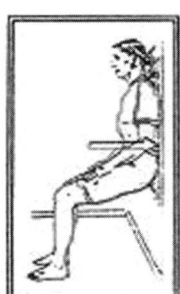

**FIG 3-26** The air bench will promote functional strength in the quadriceps.

*Source*: Reprinted with permission from *The Escogue Method of Health Through Motion*. © 1992, courtesy of HarperCollins Publishers.

# LEGS: THE BACKSWING

## LEG DRIVE—THE BACKSWING

Golfers are often told how important it is to use the legs when driving a golf ball. The benefit of leg drive is the production of long and powerful shots. However, if you ask the average golfer how to engage the legs in the golf swing, you rarely receive a definitive answer. This chapter addresses the backswing, focusing on the right-handed golfer (Figure 3-27).

> There are two components to the leg drive: The backswing, which generates the kinetic energy. The downswing, which releases the kinetic energy.

## POSITIONING THE BACK FOOT

Before your patient swings the club, the back foot must be positioned perpendicular to the imaginary Plane Line that extends from the golf ball to 10 feet behind the golf ball. When the back foot is perpendicular to the Plane Line, the hips and shoulders coil rapidly, generating kinetic energy, or power. When the back foot is "open" to the Plane Line, there is a loss of that energy, and shots hit off-line.

Have you ever wondered why, in professional baseball, the pitcher's "rubber" or step-off surface is placed parallel to the leading edge of home plate? It's done so the pitcher can place the back foot parallel to the target. During the windup, the pitcher pivots the hips and shoulders around the back leg, until they are perpendicular to the target. A right-handed pitcher keeps the weight coiled inside the back foot, and then releases the stored kinetic energy with a perpendicular push off the rubber toward home plate.

**FIG 3-27** (a) Full backswing with weight coiled inside the back foot. (b) The coil will reverse itself on the downswing by reflex, and (c) Complete transfer of the body weight to the front foot.

When swinging the golf club, the hips and shoulders start parallel to the Plane Line when addressing the ball. At the top of the backswing, the hips and shoulders are perpendicular to the Plane Line. When the back foot is perpendicular to the Plane Line, the golfer can drive his or her legs parallel to the plane line and directly toward the chosen target.

## FLEX OF THE BACK KNEE

It's important that the golfer's torso does not float upward during the backswing. The torso floats upward if the back leg straightens out during the backswing (Figure 3-30). If your patient is straightening the back leg, he or she is likely to hit just behind the ball "fat" on the downswing.

## CLINICAL EXAMINATION

**BACK FOOT:** Place a golf ball on the floor. Place a strip of tape under the ball that extends to 10 feet behind the ball. The tape is the Plane Line. Ask your patient to set up over the ball at the

**FIG 3-28** The back foot is perpendicular to the Plane Line: The hips and shoulders will turn perpendicular to the Plane Line with the weight coiled inside the back foot.

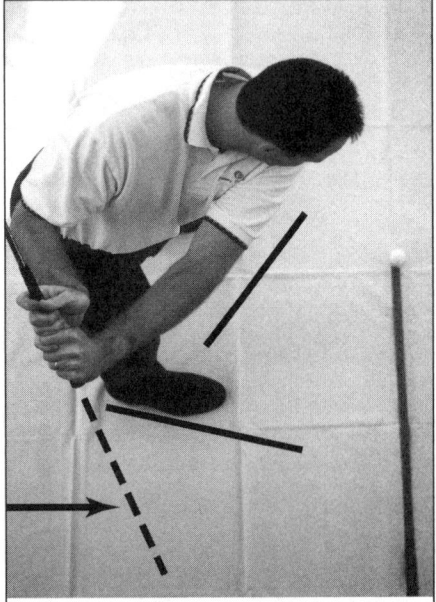

**FIG 3-29** The back foot is "open" to the Plane Line. This results in loss of power to shots hit off-line.

address position. Note the position of the back foot; it should be perpendicular to the Plane Line.

**FLEX OF RIGHT KNEE:** Have the patient swing the club to the top of the backswing, and then stop. Observe the flex of the right knee. If the knee was flexed at address, and then straightened at the top of the backswing, the torso will have elevated. Take a second look at the backswing, and watch the movement of the left foot. If the patient has to move up onto the toes, there is poor flexibility for golf in the lower leg.

## CORRECTION PROTOCOL

**OPEN THE BACK FOOT:** Ask the patient to point the back foot directly at the Plane Line. Note that the patient may have the back foot open and feel restricted in the ability to turn the hips otherwise. Check for ilium and/or sacroiliac joint physical restrictions. Another possible cause of the restricted hip turn could be that the rotators of the hip girdle are too tight. In either case, the problem can be corrected with spinal adjustments and muscle-specific stretching in the hips.

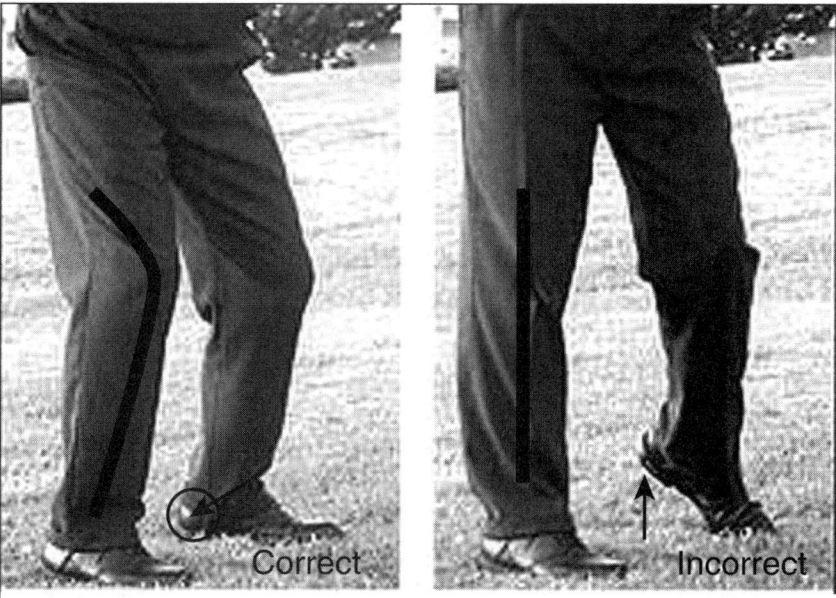

**FIG 3-30** To avoid a change of height of the torso from backswing to downswing, keep the back knee slightly flexed (left). Straightening of the back knee (right) will often result in the club striking the ground just before it hits the ball.

**Straightening the right knee:** Check the flexibility of the gastrocnemius and soleus. Have the patient work on lengthening these muscles until he or she can avoid lifting up onto the toes during the backswing (Figures 3-31 and 3-32).

# Legs: The Downswing

## Leg Drive—Downswing

The key to hitting accurate golf shots is releasing the kinetic energy of the downswing parallel to the Plane Line (Figures 3-33 and 3-34).

In the previous section, I left you at the set point: the momentary pause at the top of the backswing, just before you release all that stored kinetic energy toward the target during your downswing.

Initiation of the downswing begins when the legs drive the coiled kinetic energy directly toward the intended target. It is as if the extended left arm (of the right-handed golfer) is held motionless high above the shoulders, just for a moment, as the legs drive parallel to the Plane Line, clearing the hips so the left arm can descend the vertical plane and swing parallel to the Plane Line.

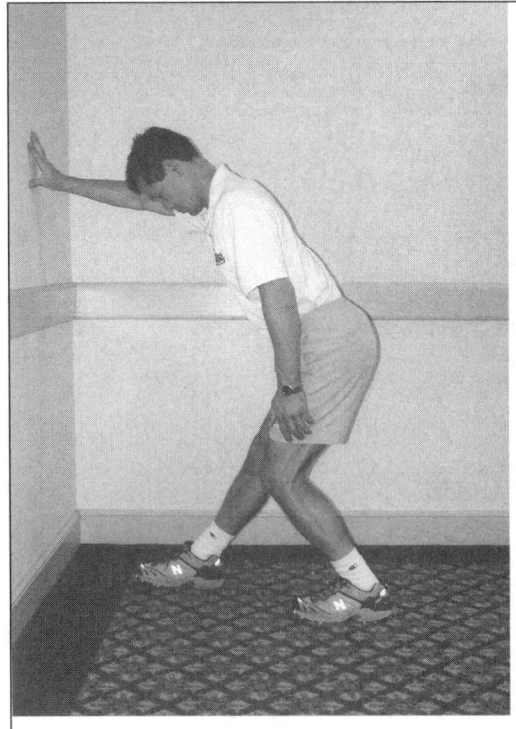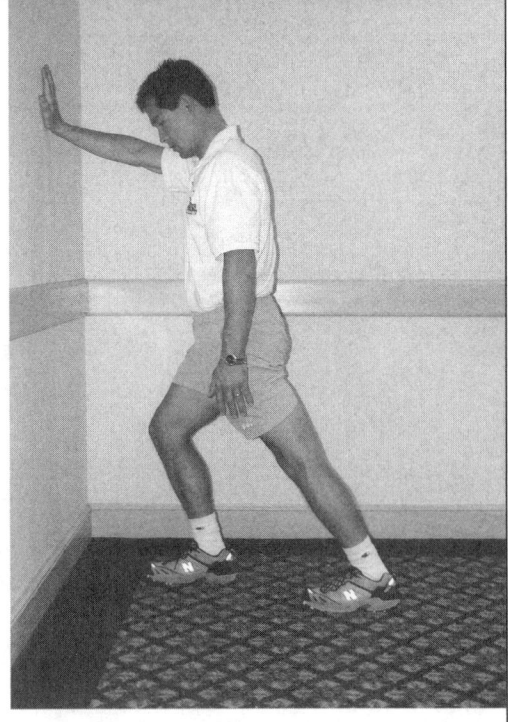

**FIGS 3-31 AND 3-32** A bilateral standing calf stretch. Hold for a minimum of one minute for each side.

## THE ADDUCTORS

Before the start of the downswing, there is a powerful 6- to 8-inch lateral slide of the pelvis as the legs drive toward the target (Figure 3-35).

Proper leg drive puts tremendous demand on the adductors.

When properly executed, the adductors stretch in response to this mechanical demand. If they are too tight, muscular strain is

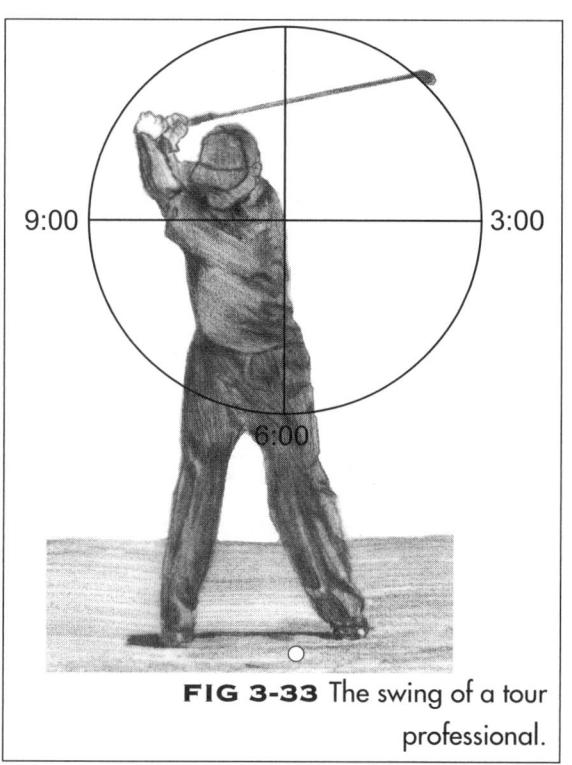

**FIG 3-33** The swing of a tour professional.

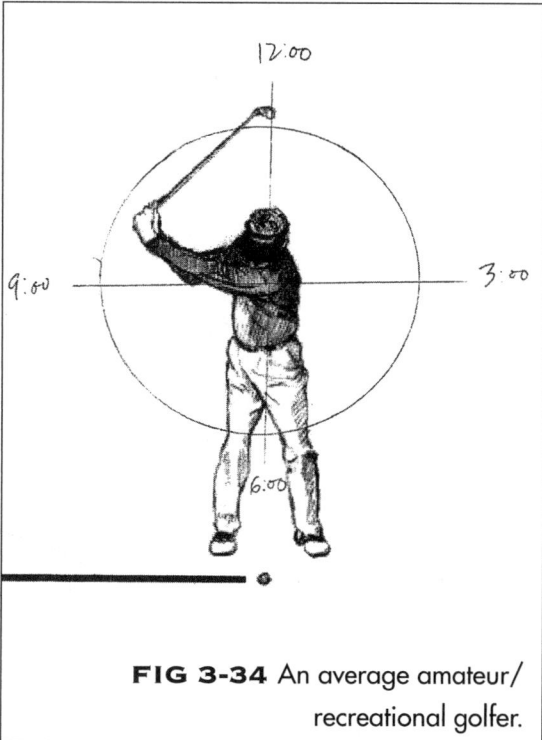

**FIG 3-34** An average amateur/ recreational golfer.

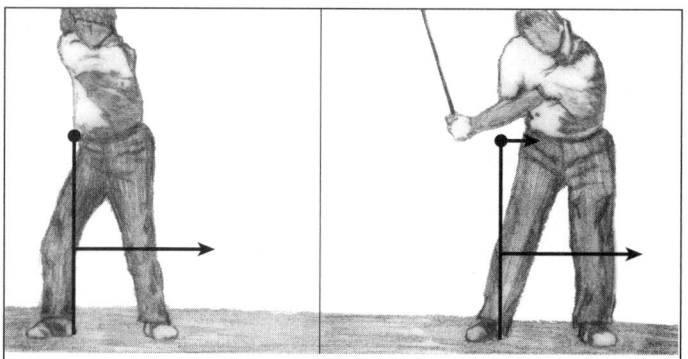

**FIG 3-35** The legs drive parallel to the Plane Line directly toward the intended target, placing mechanical demand on the adductors.

inevitable. Eventually, your patients will swing around their hips to avoid adductor strain. Swinging around tight adductors results in loss of power and contributes to an inconsistent golf swing.

**CLINICAL EXAMINATION** One way to test the flexibility of the adductors is to have your patient touch the soles of the feet while in the supine position, with the knees bent. Within 20 to 30 seconds, the knees should release to about 12 to 18 inches from the floor with no adductor discomfort. Observe whether one knee comes to rest higher than the other does. This abnormal finding indicates chronic unilateral myofascial shortening or physical restriction of the pelvic girdle.

**CORRECTION PROTOCOL** If there is physical restriction of the pelvis girdle, chiropractors can adjust accordingly. If there is poor adductor flexibility, have your patient rest in the frog position (see Part 1 and Figure 3-36) daily for 60 seconds until the adductors release.

# THE FINISH POSITION

When patients repeatedly roll their body weight over the top of their front foot, it puts the ankle at risk of injury (Figure 3-37).

After impact with ball, 100 percent of the body weight should transfer to the front foot. The front foot should be flat on the ground. For the majority of your patients, this is not the case. Instead, the body weight has rolled over the top and to the outside of the front foot.

## CORRECTION PROTOCOL

An inability to finish the swing with the weight evenly balanced on the front foot is symptomatic of being out of position during the downswing. Make certain your patients understand key points 1, 2, and 3 as discussed in Part 2, "Swing Mechanics."

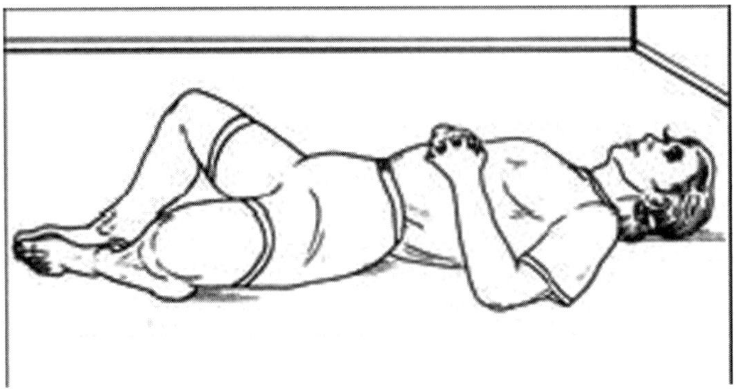

In addition, suggest your patients rehearse their finish position using the following drill. Have the patient start with the finish position and then swing *backward* to the top of the backswing.

They need to practice holding their finish position for 5 to 10 seconds at a time. Once they really feel their balance,

**FIG 3-36** The frog: a flexibility stretch for the adductors.

*Source:* Reprinted with permission from *The Escogue Method of Health Through Motion.* © 1992, courtesy of HarperCollins Publishers.

they can then swing backward to the top of the backswing. The more they practice holding the correct finish before they swing the club, the easier it becomes to "find" the correct finish position.

The following key points for the finish position are provided for your review:

> One hundred percent of the body weight should be on a flat front foot.
>
> The back foot should be positioned heel up, as if the patient is walking toward the target.
>
> The hips point toward the target.
>
> The torso may be just left of target depending on individual flexibility.
>
> Hands should be held about head high over the forward shoulder.

**FIG 3-37** Unstable finish. After impact, the body weight has rolled to the outside of the forward foot. This will put the ankle at risk of injury.

**FIG 3-38** Correct finish position.

# Knee Pain

For a stable pivot, the back foot must be positioned perpendicular to the Plane Line behind the ball. This permits the back knee to maintain it positional alignment above the ankle during the backswing.

Knee pain is common in patients who position their back foot outward relative to the Plane Line at the address position.

Patients will position their back foot open relative to the Plane Line because it hurts (because they lack the necessary flexibility) to keep the back foot perpendicular during the backswing. In short, with the back foot in the open position, patients can increase the rotation in their hip turn. Unfortunately, with the back foot open, during the downswing, the directional release of energy goes to the right of the intended target.

## Correction Protocol

The primary complaint of patients who play golf is low back pain.

Regarding patient strength and flexibility, refer to the section titled "Stablilizing the Pivot" earlier in this part. Make sure the gluteal and quadriceps muscles are conditioned to receive the weight transfer at the top of the backswing.

Regarding swing mechanics, have your patient square the back foot to the Plane Line, and then place a shallow triangular doorstop outside of the instep of the back foot. The doorstop will not permit the knee to rotate outward during the backswing.

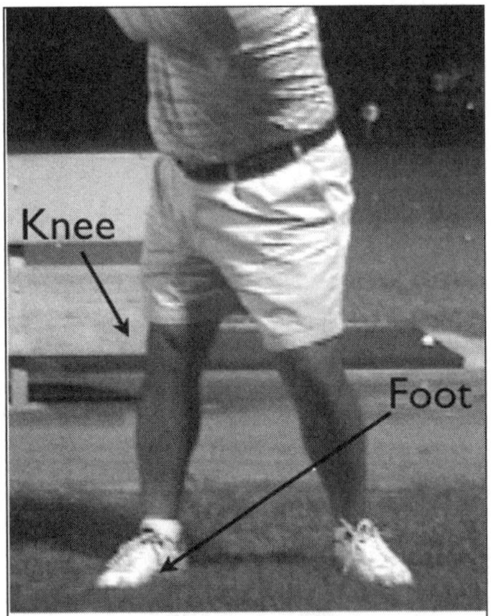

**FIG 3-39** Generally, when the back foot is positioned outward at the address position, the back knee will rotate outward during the backswing. This pivot is unstable and will put your patient at risk of knee pain and/or hip pain.

**FIG 3-40** During the backswing, the back foot must stay perpendicular to the Plane Line. When coiling, the weight must stay inside the back foot.

# HIPS

## HIP ROTATION

### Key Point 1

The back foot must be positioned perpendicular to the Plane Line (Figure 3-41).

### Key Point 2

The weight must stay inside the back foot during the backswing.

Low back pain can result from an improper pivot, lumbopelvic physical restriction, and/or inadequate flexibility. When preexisting physical problems couple with the repetitive, one-sided nature of the golf swing, they create a big problem. This explains why approximately 50 percent of patients who play golf eventually become injured and need your help.

## THE PIVOT

*Pivot* is defined as a "shaft about which related parts rotate." For the right-handed golfer, the shaft is the right leg; related parts are the hips and shoulders. There are two key points to understand about a good pivot:

The position of the golfer's back foot and the placement of the shifting body weight are the critical factors for a good pivot. The

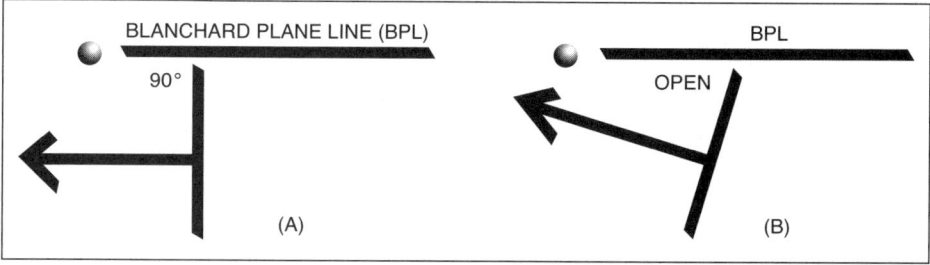

**FIG 3-41** (A) When the right foot is perpendicular to the Plane Line, the hips coil rapidly generating kinetic energy (power).
(B) The open right foot creates loss of kinetic energy and golf shots bit off-line, mostly to the right.
*Source:* Reprinted with permission from Anne Fewell.

A functional pelvic girdle is essential to maintain balance and to maximize range of motion to generate club head speed.

pivot is essential to build up kinetic energy (power) in the golf swing. Kinetic energy translates to club head speed, which is necessary to compress the golf ball. The more you compress a golf ball, the farther it goes.

## PELVIC TILT AND PELVIC ROTATION

Don't forget to check the integrity of the secondary curve of the lumbar spine.

When identified, tilt and/or rotation of the pelvic girdle impedes proper hip function during the golf swing.

Check your patient for physical restriction in the pelvic girdle.

When the pelvis becomes dysfunctional, most golfers do not position their back foot perpendicular to the Plane Line (see in Part 2, "Swing Mechanics," discussion of key point 1). When they try, often they will experience pain in the sacroiliac joint as the hips rotate during the backswing. For relief, they open the back foot to be able to rotate with less pain. Unfortunately, an open back foot results in loss of power and an inconsistent direction of ball flight.

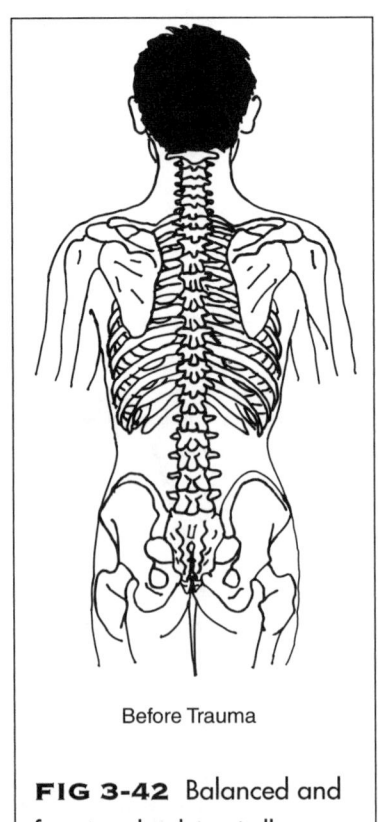

Before Trauma

**FIG 3-42** Balanced and functional pelvic girdle.

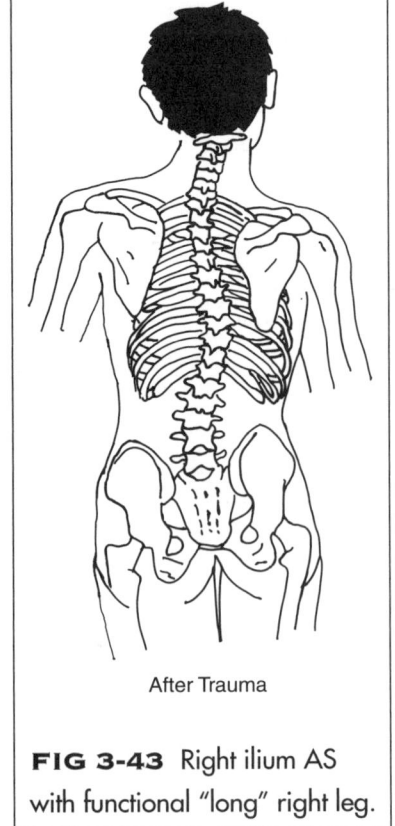

After Trauma

**FIG 3-43** Right ilium AS with functional "long" right leg.

Without adequate lumbar lordosis, your patients are at a mechanical disadvantage when they try to swing the club. The lumbar curve is an essential source of mechanical leverage for rotating the hips when swinging at a golf ball.

For the right-handed golfer, if the right ilium is subluxated anterior-superior (AS), the transfer of body weight into the right ilium during the backswing will require more effort as your patient attempts to coil into a "longer" functional leg. In addition, the coil will be restricted and painful for your patient because the right ilium will have a problem releasing to receive the transfer of weight into the right hip (Figures 3-42 and 3-43).

## POOR FLEXIBILITY

The positions shown in Figures 3-44, 3-45, and 3-46 are both diagnostic and corrective for improving the flexibility around the pelvic girdle. During examination, have your patients assume each stretch

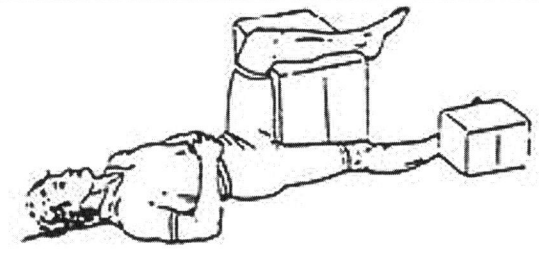

**FIG 3-44** Supine groin stretch: This stretch will facilitate release of the iliopsoas. The stretch must be done bilaterally. Once in position, squeeze the quadriceps of the down leg once every five minutes. At first, the tightest point of contraction will likely be found just about the lower two knees. When the tightest point of contraction is above the mid-thigh, switch *legs and repeat*.
*Source:* Chek, Paul, The Golf Biomechanics Manual, © 1999, C.H.E.K. Institute, Vista, CA.

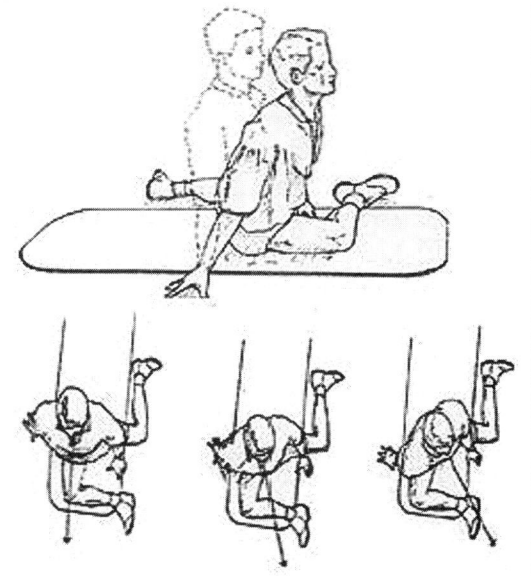

**FIG 3-45** 90/90 hip stretch: Author and consultant Paul Chek states this is the single most effective stretch a golfer can do to improve hip rotation. Hold each position for 60 seconds. Done correctly, your patients can gain improvement and strengthening of the secondary curve of the lumbar spine.
*Source:* Chek, Paul, The Golf Biomechanics Manual, © 1999, C.H.E.K. Institute, Vista, CA.

position and hold for at least 60 seconds. Make sure they keep breathing. Difficulty getting into position or labored breathing is diagnostic that this stretch is needed.

# Shoulders

## The Full Backswing: The Shoulder Turn

At least 80 percent of your patients lack the necessary shoulder and hip flexibility to swing the golf club in position through the necessary planes of motion.

**FIG 3-46** Groin stretch: Place your back to a wall, rest your forearms across the shins, and grasp the ankles. Use the forearms to open the legs to resistance, then activate the adductors to close the legs. The forearms provide resistance to closing the legs. Hold for five seconds and repeat for five repetitions.

*Source*: Chek, Paul, The Golf Biomechanics Manual, © 1999, C.H.E.K. Institute, Vista, CA.

Your patient wants to hit the ball as far as possible because it's fun. There are two basic moves to the golf swing: the backswing, which is necessary to build kinetic energy, and the downswing, the release of kinetic energy. Your patient knows a booming drive requires a long, flowing backswing. That movement requires flexibility in the shoulders and hips. It requires an unrestricted ability of the spine to bend laterally, rotate, and extend.

In less than 1.5 seconds, the golfer must generate enough kinetic energy to get the club head moving at around 100 mph. From a static posture at address, the golfer's body weight is coiled to a point inside the back foot. When the limit of the musculoligamentous elastic barrier is reached, the coil, by reflex, immediately reverses itself. The downswing starts with the hips. The golfer first slides, and then rotates toward the target to make room for the now-released kinetic energy stored in the upper torso, as the forward arm descends the vertical plane. The club head is delivered to the ball on an inside-out path through the transverse plane. As your patient stands tall, facing the intended target, there is a complete transfer of the body weight to the front foot at the finish of the swing.

But where are points of reference for the guidance of this kinetic energy? Once movement begins, how do we know if the body is moving in the proper sequence and direction? In the golf swing, we know for certain that if you wait for the feedback of the flight path of the ball, it is too late. In hitting the ball farther, flexibility of the shoulders and hips is important, but how can this be measured and improved? Adequate range of motion of your patient's shoulders includes the range of motion of the thoracic spine.

During the first phase of the backswing in a right-handed golfer, the left scapula should glide freely across the torso. Once the elastic barrier has been reached, the glenohumeral joint activates to allow continued movement of the left arm up the vertical plane. Once this elastic barrier has been reached, the thoracic spine completes the necessary rotation as the left arm climbs the vertical plane to the 11 o'clock position.

The sequence of musculoligamentous motion in the shoulders during the backswing is scapulothoracis, glenohumeral, and thoracic rotation.

> The shoulders and the torso move together, so consider them one functional unit.

## CLINICAL EXAMINATION

We need to establish a point of reference for the movement of the shoulders and torso before we begin the clinical examination. Place a golf ball on the floor. Place a piece of tape on the floor from the ball to 10 feet behind the ball. This is called the Plane Line. It is your reference point of origin for measurement of body movement and positional location of the golf club during the swing. Then, instruct your patient to complete the following steps:

1. Without a club in hand, have your patient assume the posture used to address the ball. Feet, hips, and shoulders must be parallel to the Plane Line.

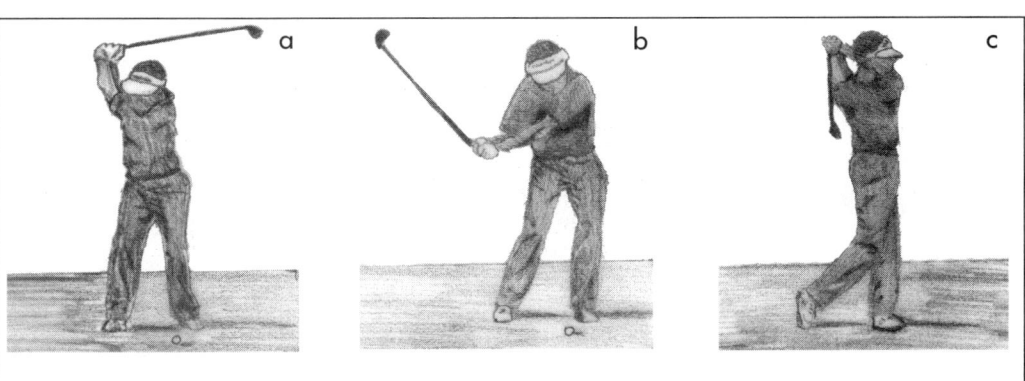

**FIG 3-47** (a) Full backswing with weight coiled inside the back foot.
(b) The coil will reverse itself on the downswing by reflex, and
(c) Complete transfer of the body weight to the front foot.

2. Place the back of the patient's right hand on the right hip pocket.

3. Extend the patient's left elbow so the left arm is pointing to the 6 o'clock position.

4. Guide your patient's left arm up the vertical plane and parallel to the plane line to the limit of the elastic barrier of the scapula as it glides outward across the upper torso. Palpate for any motion restrictions.

5. As the left arm continues to climb the vertical plane, palpate for any motion restriction at the glenohumeral joint when it engages. Palpate for any motion restriction in the thoracic spine when it engages to rotate, after the glenohumeral joint reaches the limit of its elastic barrier (Figure 3-49).

Do not be surprised at the limitation of your patient to push the left arm up the vertical plane past 9 o'clock. The average tour professional can reach to 11 o'clock without postural compromise. The biggest hitters on tour reach 12 o'clock on an imaginary clock face. It is important that your patient keep the left arm extended on the vertical plane and parallel to the plane line. Your patient should not be allowed to lift the head and torso during the exam. The goal is to try and duplicate the movement of the golf swing.

Patients know they have to take a long backswing to generate greater power and distance. But 80 percent of people lack the necessary flexibility in the scapula and glenohumeral joint, so they compensate by bending the front elbow in response to this lack of flexibility. The width of the arc collapses then, resulting in a significant loss of club head speed and power. They also pull the front arm inside, in the plane line response to lack of rotation in the thoracic spine.

Once the left arm pulls into the transverse plane during the backswing, it has to go back through the same plane during the downswing. In the golf swing, this is a prescription for disaster! Only the club head should be traveling through

> Bending the left elbow, lifting the head or torso, or movement of the left arm into the transverse plane is a compensatory move for a lack of shoulder and upper torso flexibility.

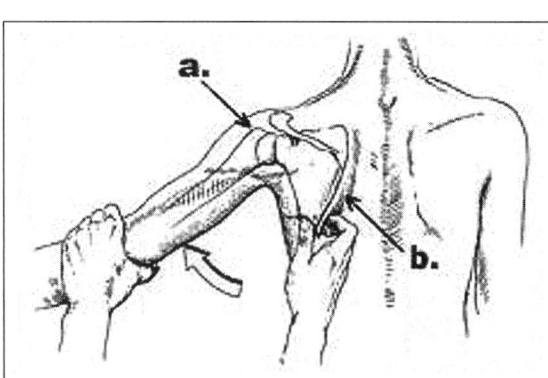

**FIG 3-48** (a) scapulothoracic (b) glenohumeral

Source: Hoppenfeld, Stanley. *Physical Examination of the Spine and Extremities*. © 1976, Appleton-Century-Crofts, Prentice-Hall, Inc.

the transverse plane, not the left arm. The left arm should move down the vertical plane (Figures 3-50 and 3-51).

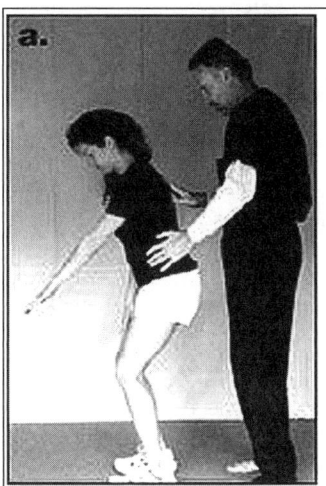

**FIG 3-49** Palpate the (a) glide of the scapula across the rib cage; (b) freedom in the glenohumeral joint; and (c) range of motion of the thoracic spine.

**FIG 3-50** Extended arm creates a wide arc.

Source: Adapted and reprinted with permission from Anne Fewell.

**FIG 3-51** Bent elbow will collapse an arc.

Source: Adapted and reprinted with permission from Anne Fewell.

## SCAPULATHORACIC RESTRICTION

**CORRECTION PROTOCOL** To increase the range of motion of the scapular glide across the rib cage, elbow curls are useful (Figure 3-52). Instruct patients to follow these steps:

1. The patient can stand or sit, with knuckles placed to the temples with palms facing out.

2. Contract the rhomboid and squeeze the scapula together. Imagine their medial borders touching.

3. Separate the scapula by hinging the forearms forward until the elbows touch in front of the chin. Do not let the wrists bend.

4. Perform three sets of 10.

## GLENOHUMERAL RESTRICTION

**CORRECTION PROTOCOL** An increase in the range of motion of this joint capsule is difficult, but the following exercise is useful (Figures 3-53 and 3-54). You may want to try this yourself to understand your patients' perspective.

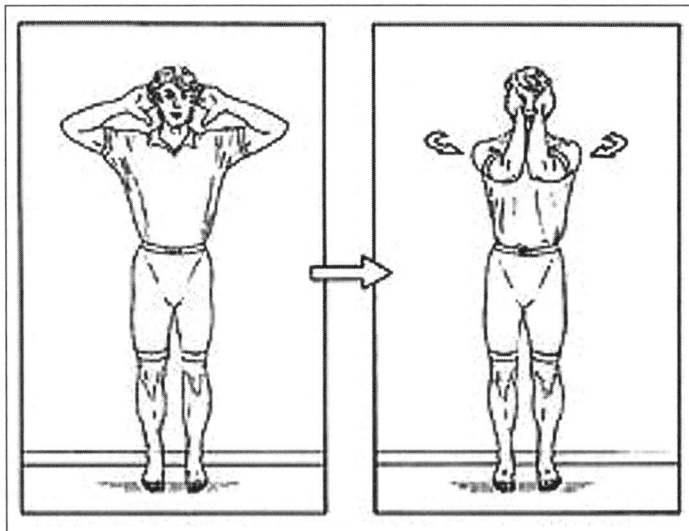

**FIG 3-52** Elbow curls will aid in scapular movement across the rib cage.

*Source:* Reprinted with permission from *The Escogue Method of Health Through Motion.* © 1992, courtesy of HarperCollins Publishers.

1. Stand at address with your feet, knees, hips, and shoulders parallel to the plane line.

2. Place the head of your driver on the plane line about 18 inches outside your back foot. Place your right thumb on the top of the grip with the back of your hand facing the target.

3. Reach back with your left hand and take hold of the grip just below your right hand. The back of your left hand should be facing just as it would during a normal golf swing.

4. Push with your right hand away from the target and pull with your left hand toward

**FIGS 3-53 AND 3-54** The glenohumeral stretch should be done bilateral to stretch the joint capsule for a better turn.

the target. Let your hips turn toward the target. Feel the pull at the glenohumeral joint.

5. Hold the stretch for 60 seconds. Remember to breathe.

## THORACIC ROTATION

**CORRECTION PROTOCOL** In addition to spinal adjustments, add the following stretch to increase thoracic range of motion (Figures 3-55 and 3-56). Your patient will need a 6-inch-round thoracic roller made of foam. Again, try this first to help you better understand the procedure.

1. Sit at one end of the roller so your entire spine and head are on top of the roller. With knees bent, spread your feet about hip-width apart for balance. Extend your arms straight up from your supine torso. Place your palms together.

2. Place your right hand over your left wrist and pull your left arm across your torso like you would during the backswing. Let your knees go left as your arms go right.

3. Feel the rotation in the thoracic spine.

4. Repeat five times each direction.

**FIGS 3-55 AND 3-56** To increase thoracic rotation, perform the stretch bilaterally.

**FIG 3-57** Wrist pain.

*Source:* © Giovanni Civardi from *Drawing Human Anatomy*. Reprinted courtesy of Sterling Publishing Co., Inc., New York, NY.

# WRIST

## WRIST PAIN

Wrist pain is common among golfers. There two general types of upper extremity injuries: "striking" injuries and repetitive strain injuries.

**THE STRIKING INJURY** This can occur when hitting the ball from the deep rough. As the club face meets the resistance of the thick grass, it decelerates rapidly relative to the arms of the golfer. In other words, the club face has all but stopped and the arms and hands keep moving at a high rate of speed. The result is a potential sprain/strain of the muscles, tendons, and ligaments of the wrist and/or elbows.

**REPETITIVE STRAIN INJURY** This syndrome can have two root causes: an incorrect grip on the club and/or a failure to keep the club shaft on the correct incline plane during the swing.

Either way, these faults have the potential to create extreme mechanical strain on the connective tissues of the upper extremities. Remember, the average patient can swing the club at 90 mph. When patients play golf or go to the driving range, they swing the club hundreds of times. If they're out of position, it's easy for them to be injured.

## THE GRIP

How your patients' hands are placed on the club has everything to do with how the wrists hinge and the right elbow folds (for the right-handed golfer) when they swing the club (Figures 3-58 and 3-59).

When your patient is out of position at the top of the backswing, it puts a mechanical strain on the right elbow and wrists when the person tries to find the downward path back to the ball with the club head traveling at 90 mph.

> When the hands are incorrectly placed on the grip, it becomes difficult to get the club shaft into the correct upward position as the club head moves behind the body.

## THE INCLINE PLANE

During the backswing, the club head moves upward, and then behind the body. Your patient is in trouble if the club shaft is not pointing at the Plane Line during the downswing. The Plane Line is the reference point for movement of the golfer's body and shaft of

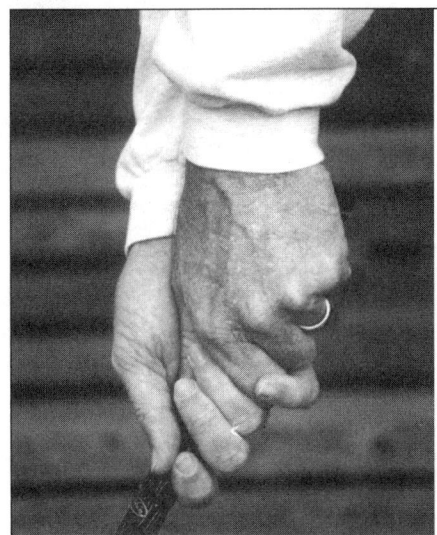
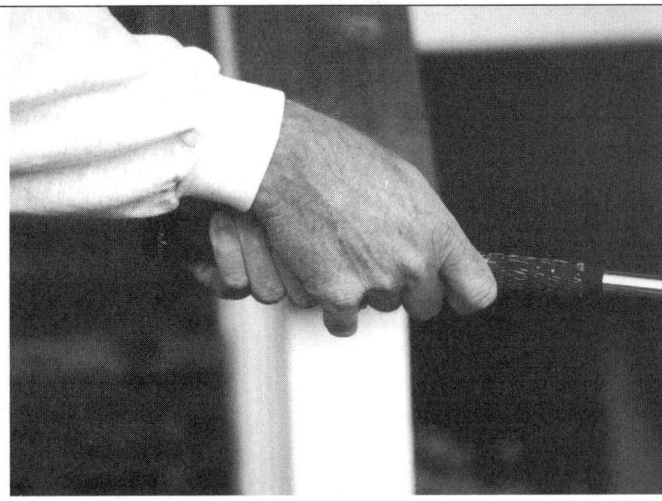

**FIGS 3-58 AND 3-59** When the hands are properly placed on the grip, the club is held firmly by only five fingers: middle, ring, and little fingers of the left hand; and middle and ring fingers of the right hand. The remaining fingers are in "light contact" for the purpose of support and "feel."

Your patient is in trouble if the club shaft is not pointing at the Plane Line during the downswing.

the golf club during the swing. The Plane Line extends from the golf ball to 10 feet behind the golf ball. When the club shaft points to the Plane Line, the club shaft is on the correct incline plane. (*Note:* There are 13 different incline planes, one for every club in the bag.)

The most common problem your patient will experience is when the incline plane of the club shaft becomes too steep or too flat during the downswing. Either way, your patient will flip, push, twist, or turn the wrists and forearms (as a last-second correction) to get the club head to make contact with the ball. This is the root cause of a repetitive strain syndrome: being out of positional alignment, and then repeatedly swinging the golf club.

**FIG 3-60** The incline plane is too flat.

**FIG 3-61** The incline plane is perfect.

**FIG 3-62** The incline plane is too steep.

## CLINICAL EXAMINATION

**THE GRIP:** It's too complicated to explain the proper grip to any patient. Instead, go to your local golf shop and buy a grip trainer for your office. Your patients' hands will fit into the molded grooves. If they complain that it feels "weird," they probably swing the club with an incorrect grip.

**THE INCLINE PLANE:** Use a swing trainer to have your patient swing the club to the top of the backswing and stop. Where is the heel light pointing relative to the plane line? If the light is shining inside the line, the incline plane is too steep. If the light is shining outside the line, the incline plane is too flat.

## CORRECTION PROTOCOL

**THE GRIP:** Advise your patients to obtain a grip trainer and take 30 to 50 practice swings daily until the corrected grip feels natural.

**INCLINE PLANE:** Have your patient use the swing trainer to take repeated practice swings with different golf clubs (sand wedge through driver). Make sure the head and heel lights from the swing

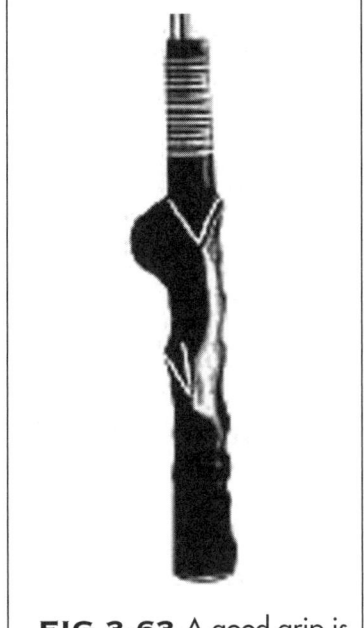

**FIG 3-63** A good grip is easily mastered by regular use of a grip trainer.

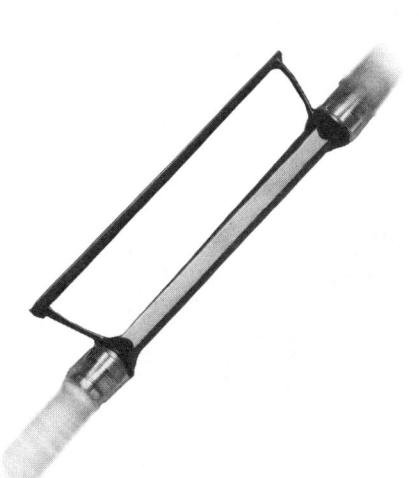

**FIG 3-64** The light beams from the Swing Trainer have the visual effect of extending the club shaft to diagnose the status of the incline plane. The lights need to point to the Plane Line to safely swing on-plane.

trainer always make contact with the plane line.

## ELBOW PAIN

Elbow pain is more common among weekend amateurs than it is among golf professionals. It is important to note that hyperextension is a consequence of flexion of the elbow joint at the top of the backswing. Flexion of the elbow joint is a consequence of poor flexibility in the thoracic spine, the scapulothoracic joint, and the glenohumeral joint.

> The primary cause of elbow pain is joint hyperextension at impact.

### CORRECTION PROTOCOL

**REGARDING PATIENT FLEXIBILITY:** See the corrective stretch protocols for the spinal rotation and the rotator cuff discussed in Part 1, "Physical Examination."

**REGARDING PATIENT SWING MECHANICS:** Have your patient place a 10-inch strip of tape across the posterior aspect of the elbow joint (mid-humerus to mid-ulna) when he or she swings the club. If your patient bends at the elbow, the tape will pull the skin and provide immediate biofeedback that the elbow is out of position.

**FIG 3-65** Secondary to centrifugal force, during the downswing, the bent elbow will "snap" and hyperextend at impact.

## FRACTURES

Fractures in golfers are uncommon, but under the conditions noted earlier, fracture of the hook of the hamate ("golfer's wrist") is the most common fracture in golf and often goes undiagnosed for several weeks to months.

The most likely cause of a fracture is a "fat" shot (even when hitting from the fairway, the club head buries into the ground behind the ball). Anytime the club head strikes a relatively heavy or immovable object, such as a rock or tree root, the golfer is at risk of fracture because the butt of the club is forced against the hypothenar region of the leading hand. Because the club head is traveling at 90 mph, breaking the hook of the hamate is a very real possibility.

Be sure to ask any patient with complaints of hand and/or wrist pain whether he or she has hit any shots that might have applied an undo amount of force to the structure in question.

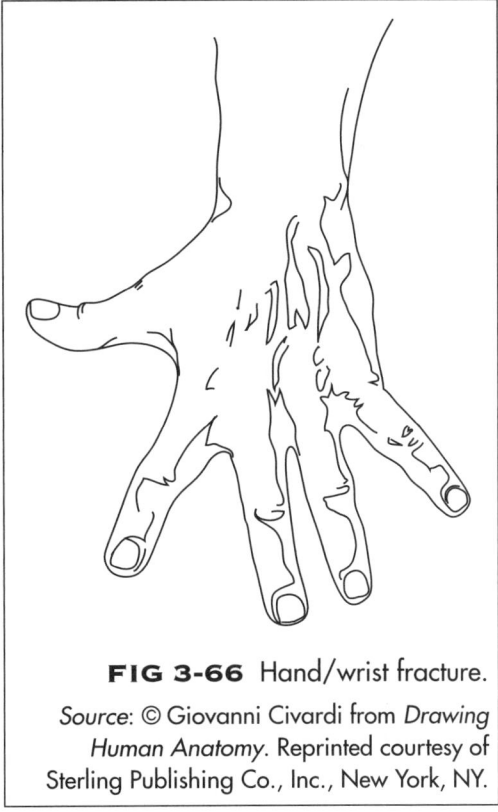

**FIG 3-66** Hand/wrist fracture.

*Source*: © Giovanni Civardi from *Drawing Human Anatomy*. Reprinted courtesy of Sterling Publishing Co., Inc., New York, NY.

# Part 2

## FOR THE PLAYER

# Chapter 4

## TRAINING FOR GOLF

### Strength, Power, and the 9-Hole Playing Lesson

# INTRODUCTION

What is the difference between "training for golf" and taking "golf lessons"?

Golf lessons are usually problem specific. Beginners can't get the ball in the air so they take lessons to learn grip, stance, and alignment to solve this problem. Experienced players take a lesson to fix a slice or to learn to hit a draw. Generally, within 30 minutes, their problem has been solved.

Training for golf implies the process of learning a skill. The skill most of your patients desire is to play well (defined as: playing at or below their established handicap on any golf course under any conditions). Unfortunately, the "process" of how to accomplish this is rarely defined.

The process is multi-dimensional. First, the body must be conditioned to swing the club safely and swing the club with power. Next, your patients need to understand their swing mechanics specific to the three key points (see Chapter 2). Finally, they need the training technology of how to take their game from the driving range to the golf course.

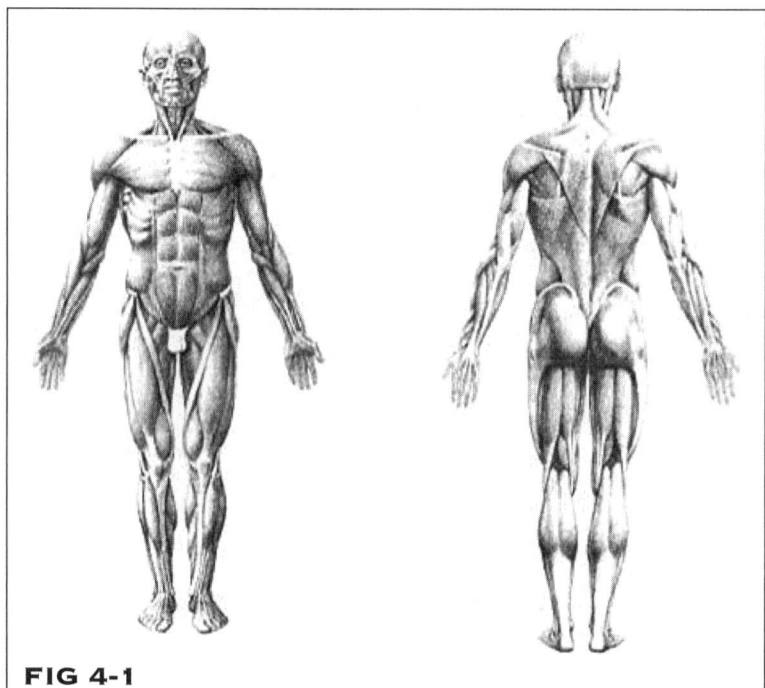

**FIG 4-1**

*Source*: © Giovanni Civardi from *Drawing Human Anatomy*. Reprinted courtesy of Sterling Publishing Co., Inc., New York, NY.

### Anterior chain

From the front of the neck, to the distal ends of each clavicle—

Working together, these muscles enable you to put your chin to your chest.

### Brachial chain

From your shoulders to the palms of your hands.

Working together, these muscles enable you to wrap your arms around, for example, another person to give a hug.

### Anterior-interior chain

From your lower spine to your upper legs.

Working together, these muscles enable you to flex at the waist and/or draw your knees to your chest.

### Posterior chain

From the back of your head to the back of your arms, then down your spine to the back of your legs and all the way to the bottom of your feet.

Working together, these muscles enable your entire body to extend backward.

## MUSCULAR CHAINS

There are 236 pairs of muscles in the human body. In 1947, Francoise Mezieres, physiotherapist and teacher of anatomy and physiology at the Paris School of Orthopedics and Massage, made a remarkable discovery: "In spite of so many different muscles, they link together into four separate chains to move the body. Inside the chains, the muscles function as a single muscle. To isolate and treat an individual muscle inside one of the four chains is useless. For improved health and function, the chains must be treated as a whole" (from www.FunctionalMuscularIntegration.com)

## THE POSTERIOR CHAIN

According to Mezieres, with proper development, the muscles inside the posterior chain will give the body "amazing strength." Athletes from Europe have been training for the Olympic Games for many, many years using Mezieres's model of functional conditioning (Figure 4-2). In short, the Mezieres model of functional conditioning is the ultimate in cross-training. Unfortunately, here in the United States, the majority of strength and fitness conditioning is done in the weight room. Too many athletes have spent too much time and energy on strengthening the anterior and brachial chains of their bodies. These athletes are very powerful in flexion activities yet very weak in activities that require extension, which engages the posterior chain.

### EXTENSION FITNESS

Golf requires extension fitness. Golf requires a healthy and functional posterior chain. If you want to hit longer and more powerful golf shots, strengthen your posterior chain. As a side benefit, as your posterior chain become functional, don't be surprised when your friends and family remark that your posture has never looked better. The following functional activities are specifically designed to increase golf-specific flexibility and to strengthen your posterior chain.

*Golf requires extension fitness.*

The following functional activities are specifically designed to increase golf-specific flexibility and strengthen your posterior chain.

## STRENGTH AND POWER

*Can you get into the right position?*

When you look at photographs of professional golfers swinging the club, you see them in positions that are out of reach for the average golfer. Remember, most professionals started swinging the golf club as a kid and never stopped. Therefore, they have been able to maintain a youthful flexibility well into their adult years.

Lots of club head speed is the key to hitting long and powerful golf shots. Club head speed is directly proportional to the width and length of your swing arc. The width and length of your swing arc is directly proportional to your flexibility.

Imagine your left arm is the short hand on an imaginary clock face, and then consider the following data about using a driver:

**FIG 4-2** Complete contraction of the posterior chain.

*Source:* ©John Lumb/Shutterstock, Inc.

1. Left arm to 9 o'clock =
   85 miles per hour (mph) of club head speed = 200 yards
   of carry

2. Left arm to 10 o'clock =
   100 mph of club head speed = 225 yards of carry

3. Left arm to 11 o'clock =
   115 mph of club head speed = 240 yards of carry

4. Left arm to 12 o'clock =
   125 mph of club head speed = 270 yards of carry

When you factor in the roll ratios, you can understand how the pros drive the ball between 265 to 295 yards. They have the flexibility that allows them to generate tremendous club head speed (Figures 4-3 and 4-4).

To widen your swing arc to generate more club head speed you need to improve your ability to flex, extend, side bend, and rotate. There may be many reasons you are not flexible. It might be you have bad posture. It might be you never stretch. It might be you have been injured and there is scar tissue. In any event, any amount of improvement in your flexibility will improve your ability to hit better-quality golf shots. Consider the stretching sequence described in the following sections as the basic minimum for golf fitness.

> To hit longer and more powerful golf shots you have to be willing to improve your flexibility.

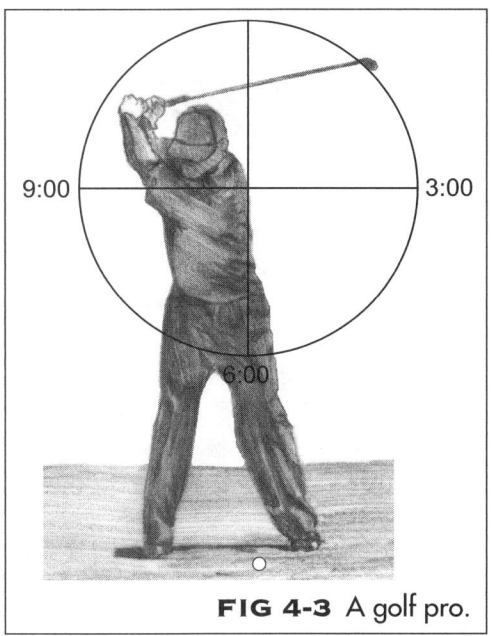

**FIG 4-3** A golf pro.

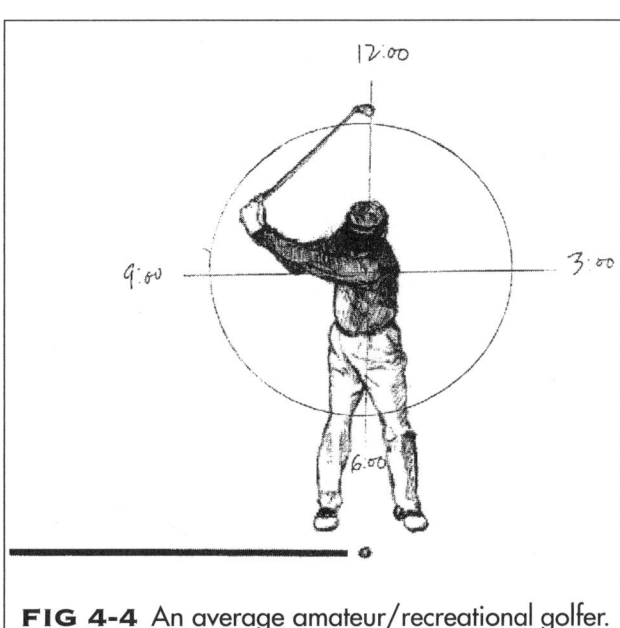

**FIG 4-4** An average amateur/recreational golfer.

## FLEXION

From a sitting position bend forward at the waist and touch your fingertips to the floor (Figure 4-5). Then, tuck your chin as you wrap your arms around your knees. Hold for 1 minute. Keep breathing.

## EXTENSION

Extension stretching should be done in the following sequence:

1. **Static Back** (from Pete Egoscue): Lie on your back (this stretch must be done on the floor), and pull your knees up directly over your hip joints (Figure 4-6). Rest your legs on a chair. Stay down for 20 minutes at a time.

2. **Supine Groin Stretch** (from Pete Egoscue): Following 20 minutes in Static Back, let one leg extend to the floor in line with your hip and shoulder (Figure 4-7). Keep the leg down for 20 minutes, then switch legs.

3. **Thoracic Roller Stretch:** Start by sitting on the floor with the thoracic roller against your spine and below your shoulder blades (Figure 4-8). Support your head with your

**Note:**
Should you experience any dizziness or blurred vision, stop stretching.

**FIG 4-5** Spinal flexion.

**FIG 4-6** Static back.

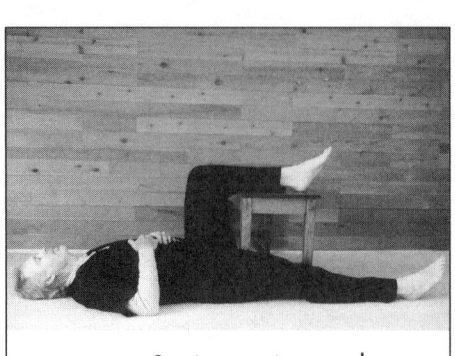

**FIG 4-7** Supine groin stretch.

**FIG 4-8** Extension with thoracic roller.

hands behind your neck. Take a deep breath. As you exhale, slowly extend your spine over the roller. Hold for 10 seconds. You can repeat the sequence at different levels of your spine.

## SPINAL CURVES

The human spine has three natural curves (Figure 4-9). The neck and lower back curves arch forward. Between your shoulder blades (where your ribs attach), the curve arches backward. Over time, your spine can mold into one forward curve. A collapse of the spinal curves weakens your spine and reduces the range of motion of your torso. This is not good. (Kapandji 1974, 20.)

The spinal curves are essential for mechanical leverage. Your muscles are stronger when they contract against the curves. In addition, with good spinal curves, in golf your torso rotates farther during the backswing. More rotation creates a wider swing arc. A wider swing arc means more centrifugal force. More centrifugal force means more club head speed to compress the ball. The more you compress the ball, the farther it travels through the air. This is *very* good.

**ROLLED TOWELS** Roll up a bath-size towel to place under your neck. When rolled it should be about 4 inches in diameter. When placed properly under your neck, the back of your head should be able to touch the floor (Figure 4-10).

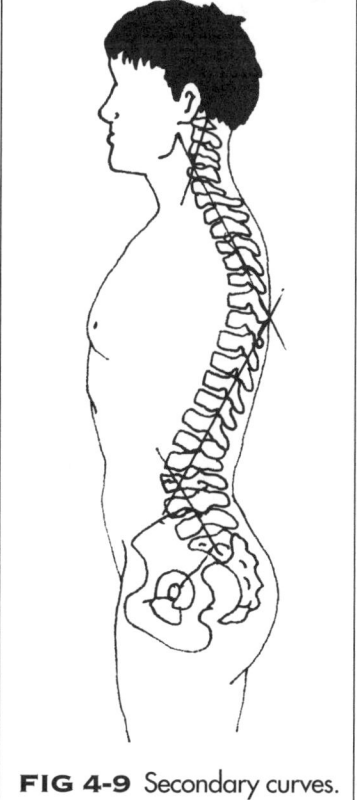

**FIG 4-9** Secondary curves.

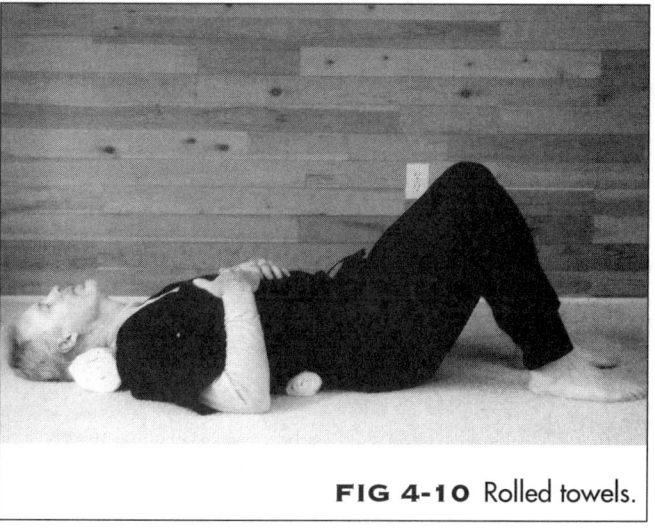

**FIG 4-10** Rolled towels.

For the lower back, use a hand towel wrapped with a wash cloth. The diameter should be about 2 inches. When properly placed, the lower-back towels are just above your belt line on your lumbar spine. Your tailbone and your rib cage should stay in contact with the floor.

### NOTE:

1. Before lying on the rolled towels, lie in Static Back for at least 5 minutes.

2. It is normal if you are uncomfortable on the towels for the first week or so. Start with 5 minutes a day on the towels. Add more time every other day until you can lay comfortably on the towels for 20 minutes.

### SIDE BEND

Use a Swiss ball for this exercise. Position yourself with one knee down on the floor. Position the Swiss ball at your side. Slowly reach over the top of the Swiss Ball until your hand touches the floor on the other side (Figure 4-11). Push yourself up on the ball, then reach over the ball with your top arm. Let your head drop. Hold for 1 minute per side. Keep breathing.

### ROTATION

Use the thoracic roller for this stretch. Lie on top of the thoracic roller lengthwise (Figure 4-12). Keep your feet apart for balance. Extend your arms up from your chest. Grab a wrist and pull your extended arm across your chest as your knees go in the opposite direction (Figure 4-13). Hold for 1 minute per side. Keep breathing.

### GIMBAL STRETCH

*Gimbal*, by definition, is the interconnection that allows one part of a mechanism to revolve independently of another revolving part. The key point to remember here is

**FIG 4-11** Side bend with swiss ball.

that at the top of the backswing, the hips revolve independently of the shoulders. Then, during the downswing, the hips release and rotate toward the target before the shoulders (Figure 4-14).

During the backswing, it is critical to keep your weight on the inside of the back foot. In fact, the rotation of your hips and shoulders is a direct response to keeping your weight on the inside of your back foot. Should you fail to do this, your hips and shoulders will drift and slide away from your target instead of winding into a tight and powerful kinetic coil.

At the top of the backswing, the hips should rotate about 45 degrees while the shoulders should rotate about 90 degrees (Figure 4-15).

> Hips rotate about 45 degrees. Shoulders rotate about 90 degrees.

**FIG 4-12** Start position.

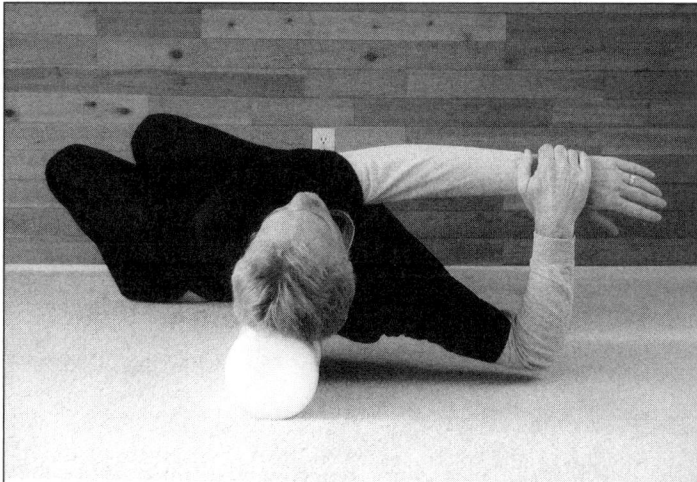

**FIG 4-13** As shoulders turn right, knees turn left (Repeat bilateral).

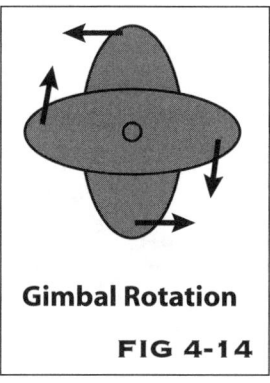

**Gimbal Rotation**

**FIG 4-14**

When done properly, the right-handed golfer should feel a significant amount of tension down the left side between the shoulder and hips.

Because the hips and shoulders are required to rotate differing amounts relative to the Plane Line behind the ball, it is important that golfers condition their body appropriately to maximize their range of motion as well as develop the propreoception necessary to maintain balance.

**CLINICAL EXAMINATION** Have your patient lie supine on the floor. Feet are up onto an exercise ball, arms extended outward with palms down (Figure 4-16). Tell the patient to, very slowly, let the knees rotate right (Figure 4-17). The right thigh should be able to touch the floor while the shoulder blades stay flat against the floor. Test bilaterally.

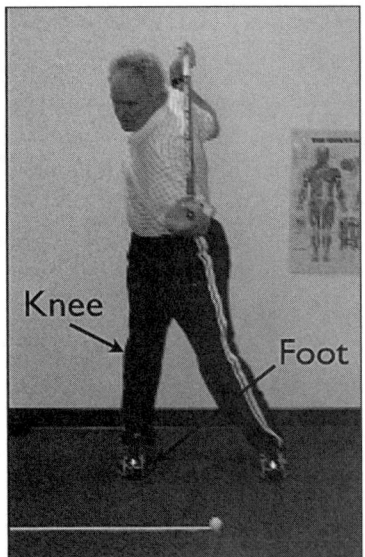

**FIG 4-15** When fully coiled at the top of the backswing, the hips and shoulders rotate independently around the stationary back leg. Note: the body weight is maintained on the inside of the back foot.

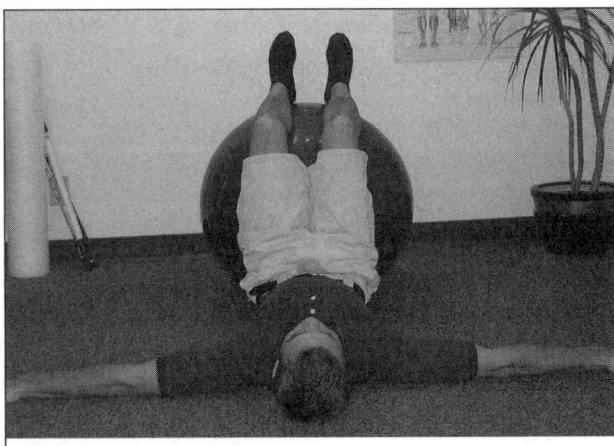

**FIG 4-16** Start position.

**FIG 4-17** Keep shoulder blades flat to floor.

**CORRECTION PROTOCOL** The examination is the correction protocol. Have the patient repeat very slowly to a tolerance of three to five times bilaterally every day until the thighs can reach the floor with the shoulder blades flat against the floor.

## THE REVERSE GIMBAL

The reverse gimbal is performed with a stabilized lumbopelvic region. The thoracic spine is balanced on top of a 36-inch-thick foam thoracic roller. The intent here is to get the extended arm parallel to the floor without rotation of the hips (Figure 4-18).

## QUADRICEPS

Strong legs are essential to playing better golf. If your quadriceps are weak, your pivot will be unstable. Golfers who walk when they play (as opposed to riding a golf cart) cover about 4 miles on a typical golf course and "backswing" their body weight onto the quadriceps about 100 times, including practice swings. Therefore, the quads need to be firm, fit, and functional to play good golf.

*Walk or, even better, run up hills or climb stairs as part of your training and conditioning regime (Figure 4-19).*

From Pete Egoscue, *The Egoscue Method of Health Through Motion* (HarperCollins 1992):

The air bench (see Part 1 for a description) promotes functional strength in your quadriceps.

Be sure to lift the toes off the ground (Figure 4-20).

Hold 3 to 5 minutes.

## LOOSENING TIGHT SHOULDER JOINTS

A capsule of connective tissue surrounds your shoulder joints that can get very tight and restrict your shoulder turn. To stretch the joint capsule follow these steps:

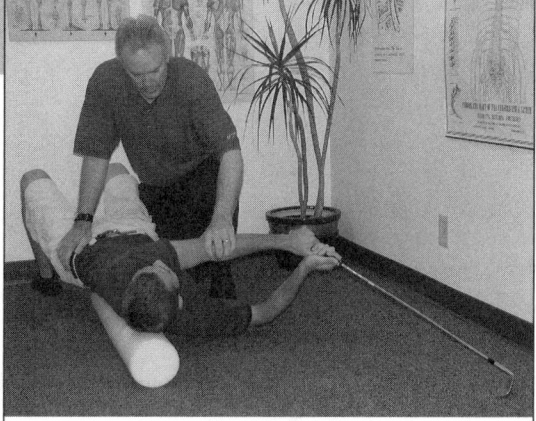

**FIG 4-18** The reverse gimbal.

1.  Stand at address with your feet, hips, and shoulders parallel to the Plane Line.

2.  Place the head of your driver on the Plane Line about 18 inches outside your right foot. Place your right thumb at the top of the grip with the back of your hand facing the target (Figure 4-21).

**FIG 4-19** Skip a step.

**FIG 4-20** Air bench (toes up).

**FIG 4-21** Start.

**FIG 4-22** Stretch for forward shoulder (Repeat bilateral).

3. Reach back with the your left hand and take hold of the grip just below the right hand. The back of your left hand should be facing just as it would during a normal golf swing.

4. Push with your right hand away from the target and pull with your left hand toward the target. Let your hips turn toward the target (Figure 4-22). Make necessary adjustments in your position until you can feel the pulling in the shoulder and rib cage under the arm.

5. Hold the stretch for 60 seconds. Don't forget to breathe.

## HANDS AND FOREARMS

Holding the correct incline plane requires strength in golfers' hands and forearms (Figures 4-23 and 4-24). Two strengthening exercises are helpful:

**GRIP BALL:** Keep a few grip balls around the house and one in the car. Squeeze firmly, alternating hands, every chance you get (Figure 4-25).

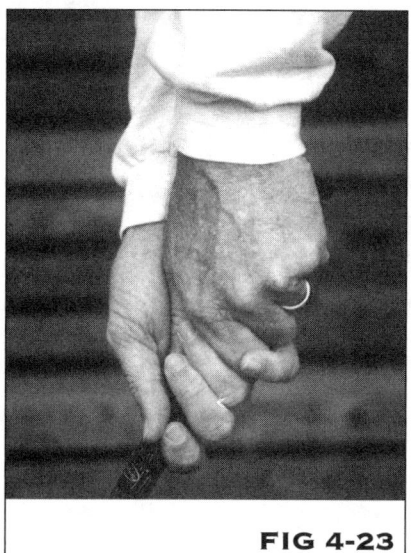

**FIG 4-23**

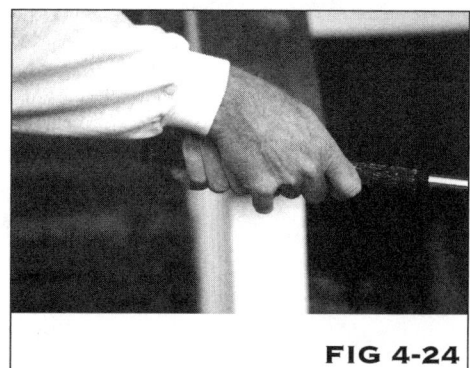

**FIG 4-24**

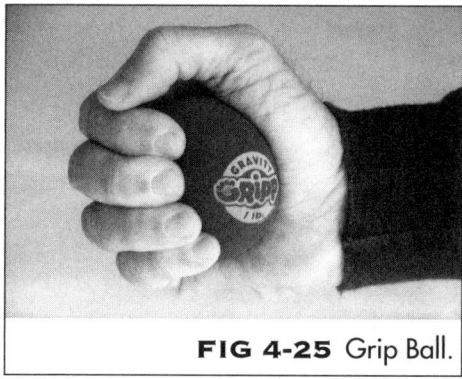

**FIG 4-25** Grip Ball.

**5-POUND HAND WEIGHT:** Holding a 5-pound barbell and keeping your forearm steady, take your wrist through a full and complete range of motion against the resistance of the weighted dumbbell (Figure 4-26). You will feel the muscles in your forearm fatigue rapidly. However, the more you do it, the stronger these muscles will become.

## CARDIOVASCULAR FITNESS

If you are physically able to walk the golf course, then walk. Sometimes arthritis or joint replacement surgery precludes walking. The primary physical benefit of walking is keeping your body's core temperature up so that your muscles can stay loose. Prolonged sitting promotes stiffness. And, from a mental perspective, it is easier to stay in a "present state of mind" when walking.

To walk the average 7,000-yard golf course is about 4 miles from tee to green. If you're hitting the ball all over the lot, then you might walk 5 miles.

To condition your body for walking the golf course, away from the golf course walk as often as possible. Using the odometer in your car, drive through your neighborhood and find several different 5-mile loops. If you walk the whole way, it will take you about 90 minutes to cover 5 miles.

*When is the last time you ran as fast as you could?*

It can be better if you alternate walking with running. Maybe after 5 minutes of walking, you can run for 1 minute. Or perhaps for you, it's better to walk for 10 minutes and run for 3 minutes. There are as many different walk/run combinations as there are people.

Running is different from jogging. Running is like being on a horse at a full gallop—it's a pretty smooth ride (Figure 4-27). Jogging is like being on a horse that's trotting; it can be a pounding. If you want to cover a quarter mile (one lap around a track), you might be better off walking the first 150 yards, running as fast as you can for 100 yards, and then walking the final 150 yards rather than slowly jogging for 400 yards. Plus, as your body conditions itself, you'll find it feels great to run fast. It's something we did almost every day as kids.

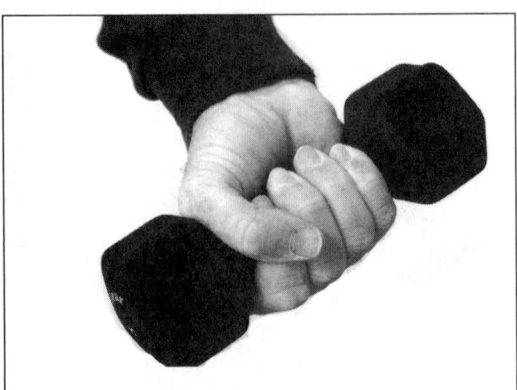

**FIG 4-26** Five pound hand weight.

## HYDRATION AND DEHYDRATION

I am going to make this very simple for you.

> If you weigh 180 pounds, drink 90 ounces of water every day.
>
> If you weigh 125 pounds, drink 62.5 ounces of water every day.
>
> Drink half the number of your body weight in ounces of water every day.

I have been treating patients for musculoskeletal injuries, aches, and pains for more than two decades. I have discovered that at least 90 percent of my patients present as chronically dehydrated. Dehydration is a root cause of nearly every ache and illness your body will experience.

The key to drinking water is to sip water all day long. Drink no more than a swallow at a time. If you must, put a spoonful of pure organic lemon juice in a 20-ounce bottle of water to make the water taste better.

*All organs and systems in your body are water-dependent.*

Soda, beer, coffee, and sports drinks all contain water but are so full of sugar they *do not count!* Understand that when you commit to drinking adequate amounts of water, you will feel awful for the first few days because your body has become addicted to the sugar in other liquids and will experience a type of withdrawal. However, in the long run, if you want to feel better, if you want to play better, drink water.

## WARM-UP FOR GOLF

Before you exercise, warm up your body. Just as you're supposed to let your car idle for a minute or two after you start it before driving to give the engine fluids a chance to circulate, you need to get your heart rate up a bit to get the blood flowing to the muscles of your body before you exercise.

Walk around the parking lot for a few minutes, do a few jumping jacks; if stairs are avail-

**FIG 4-27** Wind sprint.

able, go up and down a few times. Get your body warm and ready to move.

## PELVIC TWIST

Why? Proper alignment and unrestricted movement of the hips are essential for balance and for power.

How? Place your feet straight ahead and directly under your hips. Hold the shaft of your driver against the small of your lower back, palms facing forward (Figure 4-28). Tilt your spine slightly forward from the waist. As you slowly rotate to the left and right, keep your weight on the inside of your feet. Maintain the forward curve in your lower back. Maintain the forward tilt of your spine.

Repeat 5 to 10 times on each side.

## SHOULDER TWIST

Why? Proper alignment and unrestricted movement of the shoulders are essential for keeping your hands as far from your body as possible when you swing the club. A wide arc means power!

How? Place your feet straight ahead and directly under your hips. Hold the shaft of your driver against your upper back at the top of your shoulders blades, palms facing forward (Figure 4-29). Tilt your spine slightly forward from the waist. As you slowly rotate to the left and right, keep your weight on the inside of your feet. Stretch your spine and upper body to rotate 90 degrees left

**FIG 4-28** As you rotate, keep your weight on the inside of your feet. Maintain your forward spine angle. Repeat 5 to 10 times each side.

and right. Maintain the forward tilt of your spine (Figures 4-30 and 4-31).

Repeat 5 to 10 times on each side.

## THE LUNGE

Why? Proper alignment and unrestricted movement of the hips are essential for balance and power. The lunge stretches your hamstring and groin muscles and activates the posterior muscle chain.

How? Stand tall. Place the shaft of your driver into the small of your back, palms facing outward. Step up with one foot onto a bench or the fender of a golf cart (Figure 4-32). Push your hips

**FIGS 4-29, 4-30, 4-31** As you rotate your shoulders, keep your weight on the inside of your feet. Maintain your forward spine angle. Repeat 5 to 10 times each side.

**FIG 4-32 The Lunge** Stand up tall. Keep an arch on your lower back. Hold the stretch 30 seconds each side.

forward toward the bench. Keep the arch in your lower back. Do not bend forward at the waist.

Hold for 15 to 30 seconds on each side.

### THE TEETER-TOTTER

Why? Proper alignment and unrestricted movement of the hips are essential for balance and power. The teeter-totter will stretch your hamstring and groin muscles. Activates the posterior muscle chain.

How? Stand tall, feet together and pointing straight ahead. Hold the shaft of your driver with your hands apart, about the width of your hips, palms facing behind you. Slowly lift and extend one leg up and behind you until the leg is parallel to the ground (Figure 4-33). To maintain balance, let your arms hang directly underneath your shoulders. Try to lift your back foot higher than your hips.

Hold for 5 to 10 seconds for each leg.

### STEP-UPS

Why? Proper alignment and unrestricted movement of the hips are essential for balance and for power. The step-ups warm up your quadriceps and stretch your hamstring and groin muscles as well as activate the posterior muscle chain.

How? Stand tall. Place the shaft of your driver across the top of your shoulder blades, palms facing outward. Step up with one foot onto a bench (Figures 4-34 and 4-35). Push your hips forward

**FIG 4-33 The Teeter-Totter** Try to extend and lift the back leg higher than your hip. Hold each leg for 10 seconds.

**FIGS 4-34 AND 4-35 Step-ups** Stand tall, keep your elbows wide apart. Maintain the arch in your lower back. Step 5 times each leg.

toward the bench as you step up. Keep the arch in your lower back. Do not bend forward at the waist.

Repeat 3 to 5 times for each leg.

# FUNCTIONAL CONDITIONING FOR GOLF

## A WORD OF CAUTION

Functional conditioning should be performed only when you are warmed up and feeling well. The intent is to build explosive strength and power into your golf swing.

## LATERAL OVERS

Why? Designed to engage and strengthen the muscles of the posterior chain as well as develop functional strength and stability of the shoulders.

How? From an angle, take a running start at a bench. Plant your hands and lift your hips into the air above your shoulders. Because you approached the bench from an angle, your hips should rotate in a lateral arc over your shoulders. Maintain the arch in your lower back (Figures 4-36, 4-37, and 4-38).

Repeat 3 to 5 times each direction.

## DONKEY KICKS

Why? Designed to engage and strengthen the muscles of the posterior chain as well as develop functional strength and stability of the shoulders.

**FIGS 4-36, 4-37, AND 4-38 Lateral Overs** Try and lift your hips higher than your shoulders. Arch your lower back (5 times each way).

How? From straight away, take a running start at bench. Plant your hands about shoulder width apart, lift your hips into the air above your shoulders, and then kick your heels into the air (Figures 4-39 and 4-40). Maintain the arch in your lower back.

Repeat 5 to 10 times.

## LATERAL UNDERS

Why? This exercise is designed to engage and strengthen the muscles of the posterior chain as well as stretch your hamstrings and develop functional strength and stability of your quadriceps.

How? Approach an overhanging bar from the side. Slide your body under the bar with a sidestep. Keep your head up and keep an arch in your lower back (Figures 4-41 and 4-42). As the bar gets lower, the exercise gets harder.

Repeat 3 to 5 times each way.

**FIGS 4-39 AND 4-40 Donkey Kicks** Try and lift your hips above your shoulders, then kick upward into a hand-stand (10 times).

**FIGS 4-41 AND 4-42 Lateral Unders** Keep your head up and an arch in your lower back (5 times each way).

## STRAIGHT UNDERS

Why? This exercise is designed to engage and strengthen the muscles of the posterior chain as well as stretch your hamstrings and develop functional strength and stability of your quadriceps.

How? Approach an overhanging bar straight on. Slide your body under the bar by stepping forward. Keep your head up and keep an arch in your lower back (Figure 4-43). As the bar gets lower, the exercise gets harder.

Repeat 4 to 6 times and alternate your leading step.

## INCH WORM

Why? This exercise is designed to engage and strengthen the muscles of the posterior chain as well as develop functional strength and stability of your shoulders and abdominal muscles.

How? Start by standing tall, feet under your hips and pointing straight ahead. Bend at the waist to touch your hands to the ground (Figure 4-44). Walk out with your hands until you are in a push-up position (Figures 4-45 and 4-46). Then, with your hands stationary, walk your feet toward your hands (Figure 4-47).

Repeat 3 to 5 times.

## ROTATOR CUFF

Why? This exercise develops functional stability and strength of the muscles in the shoulder as well as engages the muscles of the posterior chain.

**FIG 4-43 Straight Unders** Keep your head up. Keep the arch in your lower back (10 times).

**FIGS 4-44, 4-45, 4-46, AND 4-47 Inch Worm** Start with feet stationary, then walk out on your hands until you reach the push-up position. Then keep hands stationary and walk your feet up to your hands (5 times).

131

How? Start in a three-point stance with your down hand underneath your shoulder and feet apart the width of your hips (Figure 4-48). Hold a 3- to 5-pound dumbbell in your free hand. Then, rotate and open to a full 90-degree turn with your shoulders. Try to extend and reach to the sky as far as you can (Figure 4-49).

Repeat 3 to 5 times each side.

## TRAINING FOR GOLF

### AT HOME OR OFFICE

You don't have to play 18 holes of golf or go to the driving range to be training for golf. There is plenty you can do at your home or office.

To learn a new motor skill takes about 21 consecutive days of practice. Each day you must repeat the motion or movement 30 to 50 times, as perfectly as possible. After 21 days, your new skill will have become second nature. You no longer have to give it conscious thought. It's just like learning to drive your car. In the beginning, driving took a lot of conscious effort and energy to remember so many different rules and skills at one time, but then it became second nature. And so will the new skills you practice for golf.

> Just like driving a car: after a while, you quit thinking and started doing, by unconscious reflex.

### THE SWING LIGHT TRAINER

Use a Swing Light Trainer every day for 5 to 10 minutes at your home or office. The Swing Light Trainer was created to meet the stated needs for motor development (Figure 4-50). When held to the grip of any club, the Light Trainer gives you immediate feedback on whether you are out of position during your swing. The very moment the lights lose contact with the Plane Line behind the ball, stop! Do not continue. Start over again until your rehearsal is near perfect every time. Remember never to swing the same club more than twice. By constantly changing clubs as you rehearse, you are training yourself to play golf the way the game is played out on the course.

**FIGS 4-48 AND 4-49 Rotator Cuff** Start with 3-point stance. Back straight. Hold a 3-5 lb. dumbbell in your free arm. Rotate and open to a full 90 degree turn with your shoulders. Reach to the sky (5 times each side).

## THE DRIVING RANGE

The driving range offers an excellent opportunity to train for golf. You need to learn how to condition your body and play golf at the range instead of simply hitting balls at the range.

Why not get into the mind-set of cross-training your body at the driving range. So, instead of hitting balls for an hour and then going to the gym for an hour a few days a week, combine your time and efforts. At the range, use the first 10 to 15 minutes of your time to stretch and strengthen your body, just like you would if you went to the gym. Spend the next 30 minutes playing "range golf." Finally, take another 20 to 30 minutes to do some cardiovascular and functional conditioning for your body. Imagine, when asked if you're going to go play golf again, you could reply, "No, I'm going to work out."

> At the range, you have an opportunity to train your body and your mind.

## BEFORE YOU EVER HIT A BALL

I recommend you practice swinging three different clubs a minimum of 10 swings each before you strike your first ball at the range (the driver, a 5-iron, and a 9-iron make up a good spread). Never swing the same club more than 2 or 3 times before changing clubs. This should take at least 10 to 15 minutes if you do it right. With each club, pay particular attention to your three key points:

### Key Point 1

Keep your back foot perpendicular to the Plane Line (Figure 4-52).

### Key Point 2

Make certain you are sweeping your forward arm parallel to the Plane Line (Figure 4-53).

### Key Point 3

Make certain your club shaft points to the Plane Line (Figure 4-54).

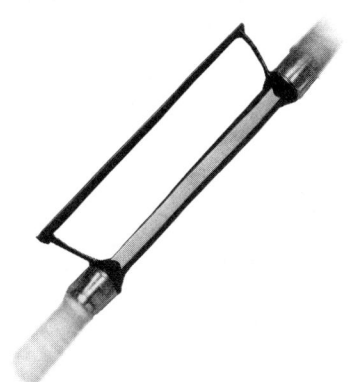

**FIG 4-50** The Swing Light Trainer.

Repeat this same training sequence with your 5-iron, and then again with your 9-iron. Remember, you need to get used to swinging the club on different incline planes. In addition to warming your muscles and joints, you are training yourself to re-create the

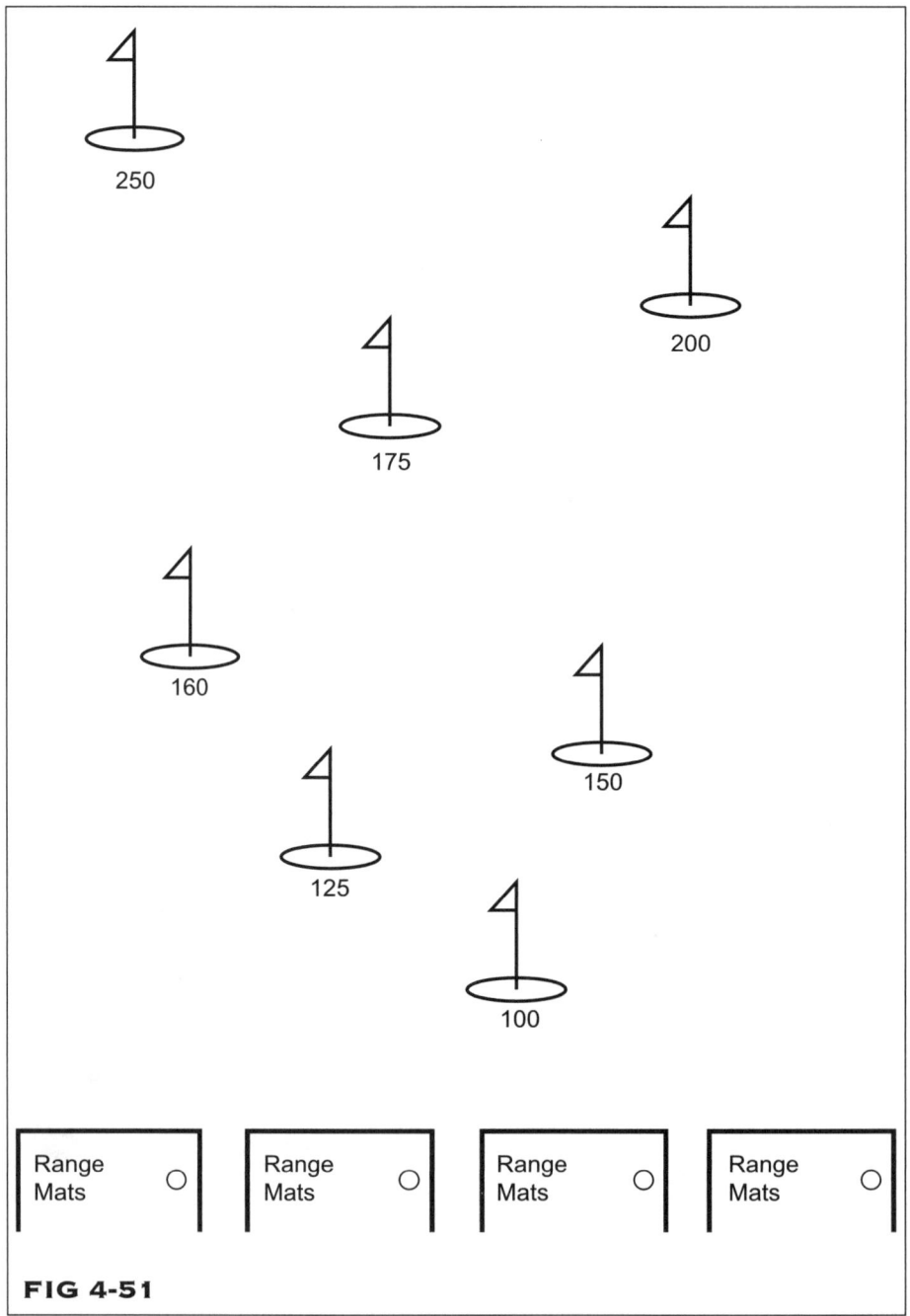

**FIG 4-51**

rehearsal swings you have been doing at your home and office, only now without the Swing Light Trainer.

## PICK YOUR TARGET

You should never hit a ball without having a specific target in mind. Why? Because the goal in golf is not to hit the ball, but to hit the target with the ball (Figure 4-55).

Take a few practice swings to see if you can hold the image of the target in your mind throughout your entire swing. If you find your thoughts drifting to the swing mechanics of your body or to the ball itself, please read Chapter 5 "Fear and Freedom: A Golfer's Guide to Mental Health."

> You should never hit a ball without having a specific target in mind. Remember: The goal in golf is not to hit the ball. The goal is to hit the target with the ball.

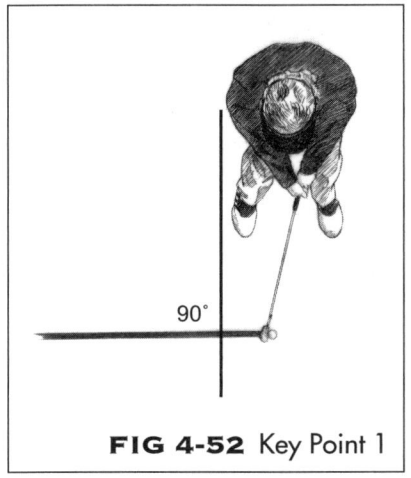

**FIG 4-52** Key Point 1

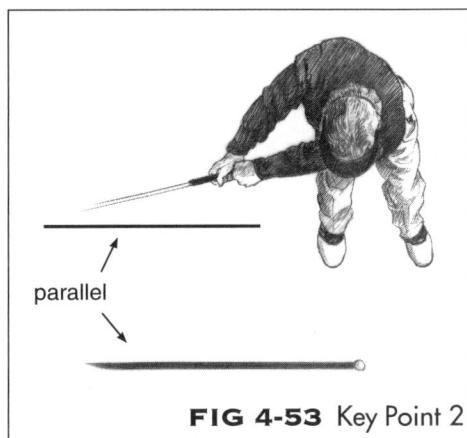

**FIG 4-53** Key Point 2

**FIG 4-54** Key Point 3

## TRAINING SEQUENCE

You have warmed up by stretching and by swinging the club. You have rehearsed your three key points. You're finally ready to hit some balls. Start with your 9-iron. Pick the target on the range that fits best with your 9-iron, go through your preshot routine, and

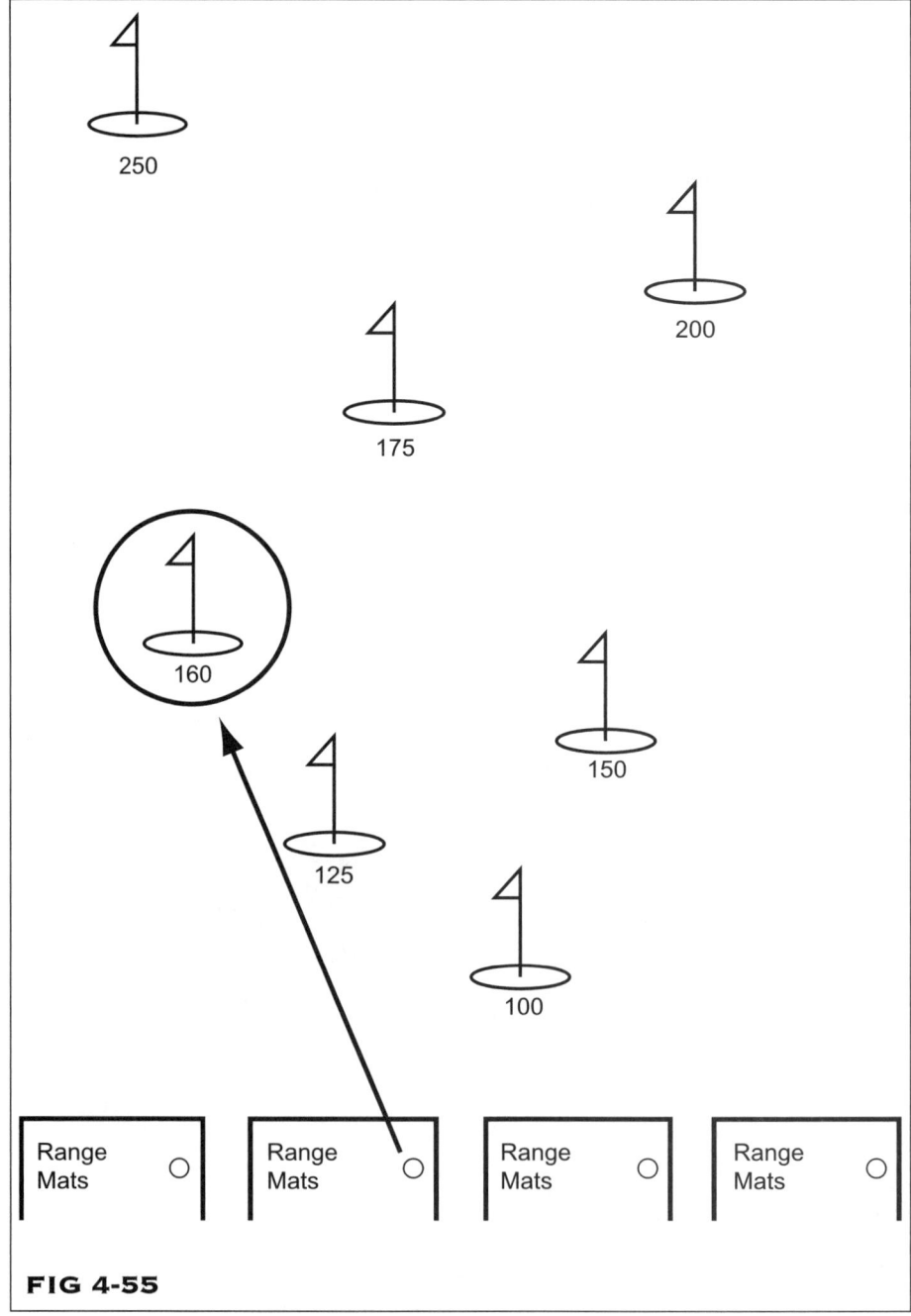

**FIG 4-55**

swing away. One swing, that's it! Pick a different club, perhaps the driver. Pick your target, go through your preshot routine, and swing away. One swing, that's it! Pick a different club, perhaps your 5-iron. Pick your target, go through your preshot routine, and swing away. One swing, that's it!

Direction—19/36 Distance—22/36

Rate each shot by its direction and distance. After 36 full-swing shots (that's like playing 18 holes), your efforts might be recorded like this:

**REGARDING DIRECTION:** 19 shots of the 36 shots actually went through the air, pretty much on-line to the target. The other shots went "elsewhere."

**REGARDING DISTANCE:** 22 shots of the 36 shots actually went through the air the intended distance. The other shots fell either way short or way long (Figure 4-56).

You are constantly changing targets, constantly changing clubs, and constantly keeping "score" of your efforts. When you get better stats at the driving range, you will get better stats on the golf course. Better stats on the gold course means lower scores!

> By now you might have realized, your training sequence is more like playing golf than hitting ball after ball is.

> You should only have two concerns after every swing.
>
> First, did you hold the image of the target in your mind from start to finish?
>
> Second, are you satisfied with the result (Figure 4-57)?

If you can respond yes and yes, pick another target and another club. If you decide to follow this training protocol, you should not have to hit more and 30 to 45 balls.

If you were not satisfied with the result, go back to rehearsal of your three key points without hitting a ball. I guarantee you will find one or more of the three key points are at the root of the cause. If necessary, invest in a video camera and tape your golf swings. Check and recheck your three key points.

> If you could not hold the target in your mind during the swing, go back to rehearsing swings at the target without hitting the ball.

## CARDIOVASCULAR AND FUNCTIONAL TRAINING

To finish your workout, why not run or at least take a very brisk walk for 15 to 30 minutes. It would be even better if you could find some hills or stairs to climb up and down. If there was a park nearby, you could use the equipment to do your functional conditioning by climbing over and under bars and benches.

# THE 9-HOLE PLAYING LESSON

> Regardless of your level of ability, there are four key statistics you need to manage to play better golf.

## FAIRWAYS IN REGULATION

*In regulation* means "relative to par." If you cannot shoot par, this stat becomes meaningless, for now. However, this is a very important stat.

A great way to train for this stat is to play for the fairway yardage markers off the tee box. Start your training with the 200-yard markers. If you are playing a 375-yard par 4, it's 175 yards to the 200-yard marker. The next hole might be a 338-yard par 4, so it's 138 yards to the 200-yard marker. On a 442-yard par 4, it's 242 yards to the 200-yard marker (Figure 4-58). And so it goes.

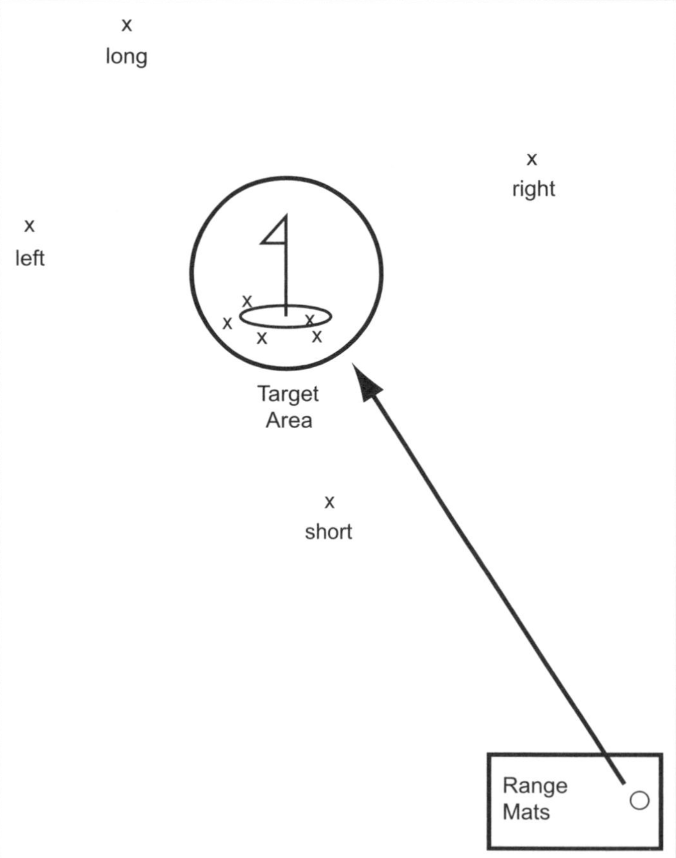

If you will take the time and train yourself to actually hit 70 percent of the par 4 fairways from the tee box to the 200-yard markers for five consecutive rounds, you are truly gaining on playing better golf.

Next, progress to playing from the tee box to the 175-yard markers. If you're playing a 365-yard par 4, it's a 190-yard tee shot. Again, take the time to train yourself to actually hit 70 percent of the par 4 fairways from the tee box to the 175-yard marker before progressing to the next level.

Once your training progresses to where you are consistently playing from the tee box to the 150-yard markers, you are likely hitting driver and doing quite well!

**FIG 4-56**

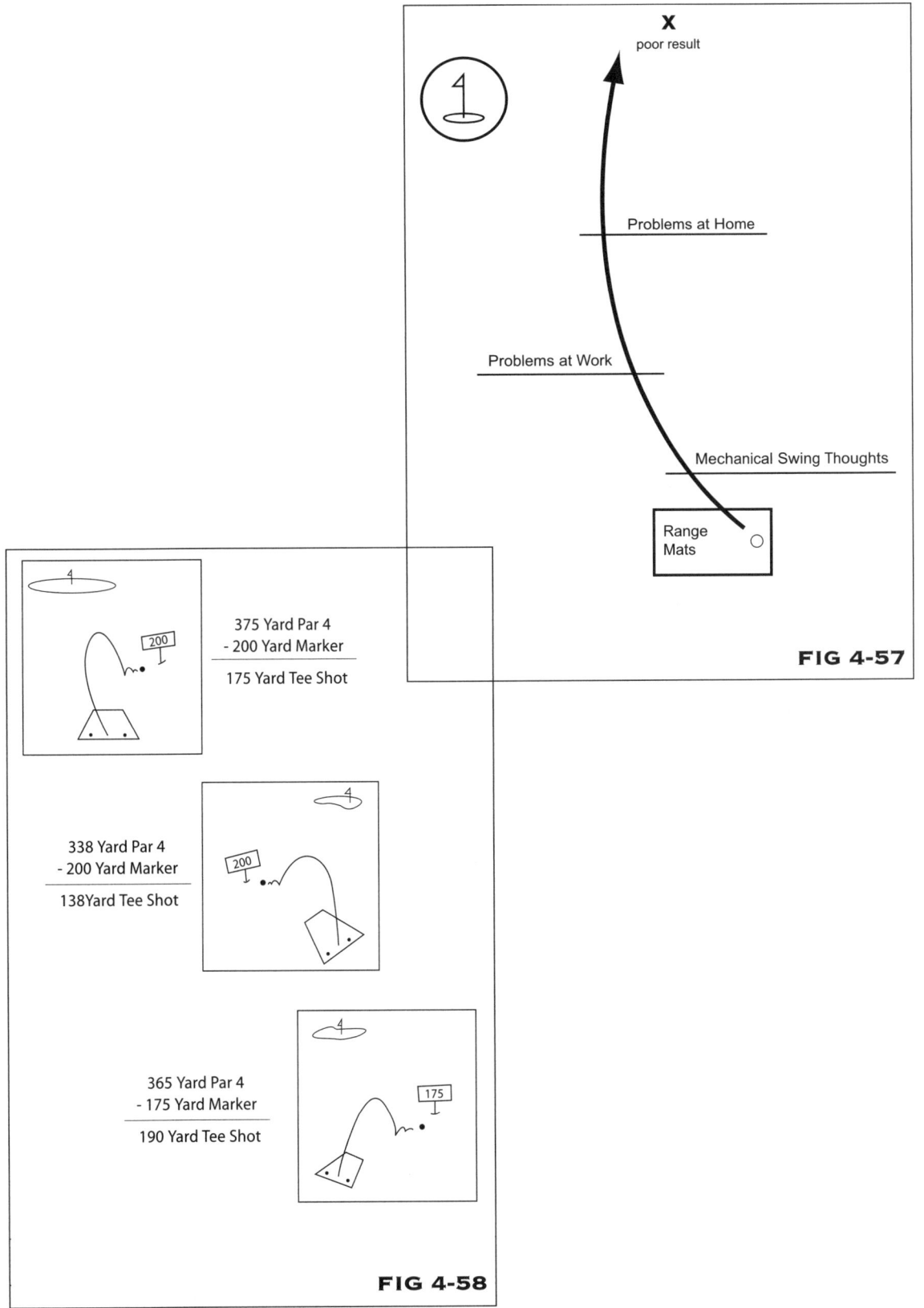

**X**
poor result

Problems at Home

Problems at Work

Mechanical Swing Thoughts

Range
Mats

**FIG 4-57**

375 Yard Par 4
- 200 Yard Marker

175 Yard Tee Shot

338 Yard Par 4
- 200 Yard Marker

138Yard Tee Shot

365 Yard Par 4
- 175 Yard Marker

190 Yard Tee Shot

**FIG 4-58**

Remember: You are no longer playing for score. You are now training to score better.

The benefit of this training regime is two-fold: you are learning to control the direction of your shots, and it's great practice for distance control. You can never tell how far you've hit the ball on the driving range because you can't walk it off.

The key to better play is better stats (Figure 4-59).

This is not going to be easy. There will be no instant gratification. However, you have found your starting place. If you can commit to this, you will improve and you will stay improved.

## GREENS IN REGULATION

*In regulation* means "relative to par." If you cannot shoot par, this stat becomes meaningless, for now. However, this is also a very important stat.

Training for this stat requires you to work from the inside out. Start inside the 100-yard marker in the fairway and progress outward to the 200-yard marker in the fairway.

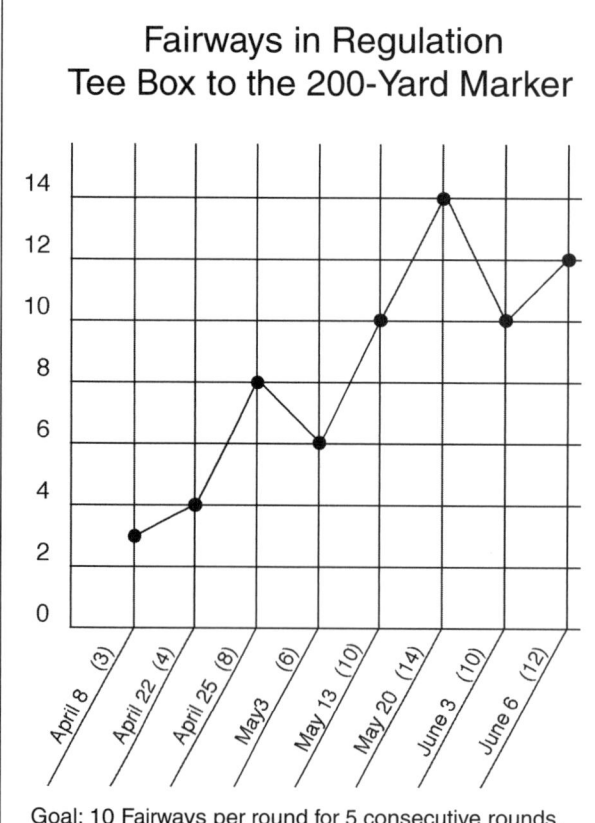

Fairways in Regulation
Tee Box to the 200-Yard Marker

Goal: 10 Fairways per round for 5 consecutive rounds.

**FIG 4-59**

**EXAMPLE 1:** You are playing the 200-yard markers from the tee box. Your ball stops at your mark in the fairway. You're now 200 yards from the green. Your next shot will be aimed at the 100-yard marker in the fairway. Your shot rolls to a sprinkler head 92 yards from the green. Now your job is to hit the green in regulation from inside the 100-yard marker in the fairway (Figure 4-60).

**EXAMPLE 2:** You're playing the 200-yard markers off the tee box. You fan the tee ball off-line and it comes to rest in the right rough 225 yards from the green. Your next shot will be 125 yards to hit the 100-yard marker in the fairway. Now your job is to hit the green in regulation from the 100-yard marker.

Once you can hit the green 70 percent of the time from the 100-yard marker in the fairway, progress outward from the 125-yard marker.

You have to be willing to progress in steps. You have to stay inside the training parameters you have established for yourself until you make the stat. Making the stat implies 70 percent of the time for a minimum of five consecutive rounds of golf.

Commit to training and eventually you will progress to hitting the green from the 200-yard markers 70 percent of the time.

## GETTING UP AND DOWN

Simply stated, when you miss the green, you take one shot to get the ball on the green and take one putt to finish the hole.

Short-game guru Dave Pelz says, "Picking the correct club for shots around the green is as much a matter of mathematics as it is feel. When you apply the proper technique and force your choice of club will come down to a simple formula." (From Dave Pelz, *Short Game Bible* [Broadway Books, 1999]). This is because every club has a predictable carry to roll ratio.

### MEMORIZE THESE CHIPPING RATIOS:

Divide Roll by Carry and call it "X". Subtract "X" from 12. The answer is your iron of choice (Roll ÷ Carry = X 12 − X = Iron of choice).

S-iron 1 to 1

P-iron 1 to 2

9-iron 1 to 3

8-iron 1 to 4

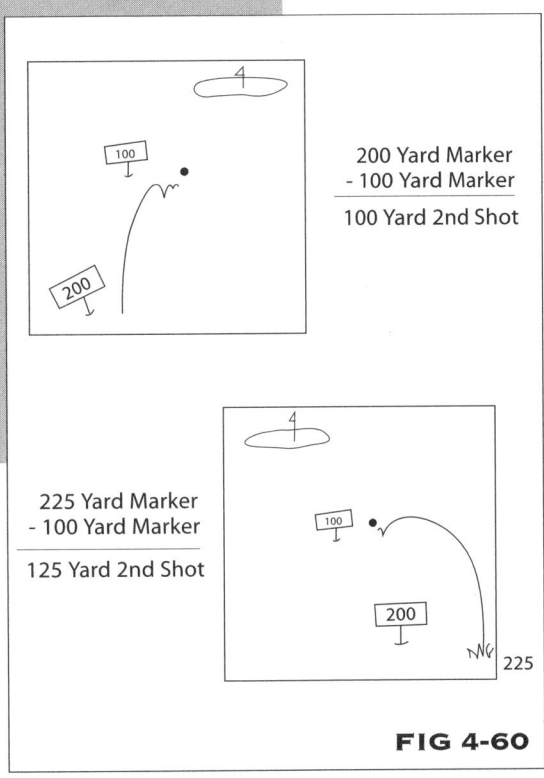

200 Yard Marker - 100 Yard Marker

100 Yard 2nd Shot

225 Yard Marker - 100 Yard Marker

125 Yard 2nd Shot

**FIG 4-60**

These ratios mean that for every foot of carry through the air, the ball will roll a predictable distance on the green based on the loft of the club. Use these ratios to step off the distance from your ball to the best landing spot on the green. Then, step off the distance from your landing spot to the cup. Knowing how far the ball will roll when you hit your landing spot is very important if you intend to improve your up and down stat.

For example, your ball has come to rest in the light rough 32 feet from the cup. For your ball to reach the putting surface, it must carry 8 feet through the air, and then roll 24 feet to the cup (Figure 4-61). In other words, for every foot of carry, the ball must roll 3 feet. This is a ratio of 1 to 3. When you reference the preceding chart, the 9-iron has a carry to roll ratio of 1 to 3. Hit your landing spot with the 9-iron, and the ball will release and roll 24 feet.

Adjustments are always necessary for slope and speed of green, but knowing your carry to roll ratios is your first essential step toward reaching your goal of getting up and down 70 percent of the time whenever you miss the green.

The next challenge in getting up and down is hitting your landing zone on the green. There is a great way for you to train yourself to do this, as described here.

When you are forced to wait at the tee box because of slow play, take out your sand wedge, pick out different targets such as a leaf or a brown spot on the grass from 5 to 15 feet out, and work on landing your shot on the selected mark.

During a typical 18 holes of golf, you will have a lot of time to work on similar shots around the tee box with a pitching wedge and

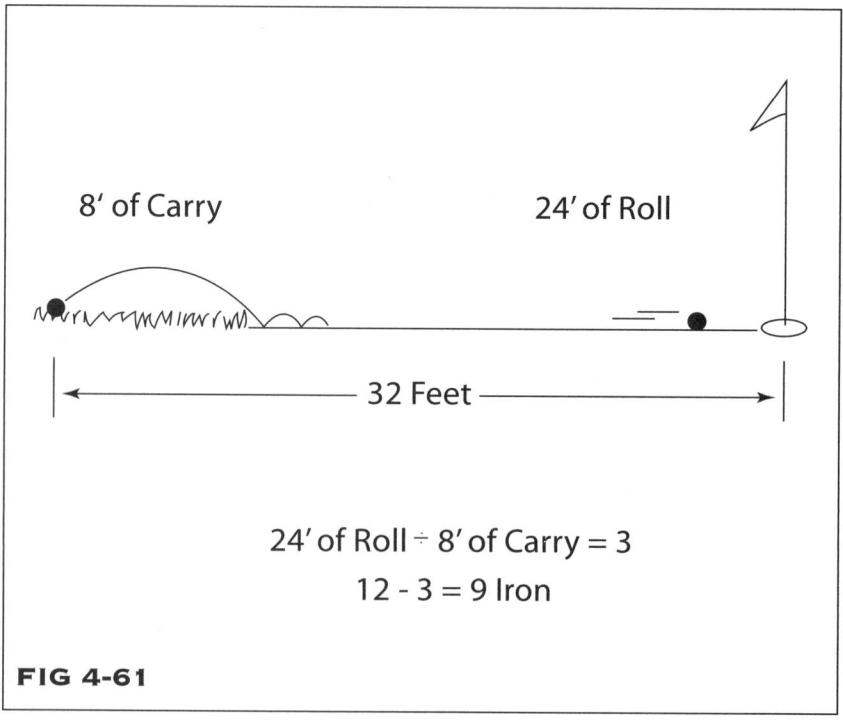

8' of Carry          24' of Roll

|←———————— 32 Feet ————————→|

24' of Roll ÷ 8' of Carry = 3
12 - 3 = 9 Iron

**FIG 4-61**

a 9-iron. These "touch shots" require hours of practice. You are not taking divots. You're just sweeping the ball out of all types of different lies in the grass on and around the tee box.

At first, you will be amazed at how hard it is actually to hit your target spots. In time, you can hit your spot from different kinds of lies with a varied trajectory. You can float it in with a soft flop or drive it in with lots of backspin.

The better you can get with your wedges around the green, the better your chance of getting up and down.

**SAND SAVES** A word about getting up and down from sand traps. You would think that hitting hundreds of balls out of sand traps would be the best way to learn to get out of sand traps. Maybe this is true, but I did not improve my bunker play until I started hitting flop shots from the ball sitting down in an inch or two of grass (Figures 4-62 and 4-63). I noticed how similar this type of shot felt to a well-struck bunker shot.

When you train hitting flop shots from deep grass and you can feel the bounce of the club head off the underlying earth, you are on your way to improvement. Duplication of this feeling in the bunker produces better explosion shots from the sand.

## PUTTS PER ROUND

You might think that two-putting every green would be a great stat. And it would if you hit every green in regulation. However, even world-class golf professionals miss 3 of every 10 greens they shoot for. Your goal in putting is to average about 30 putts per round of golf.

Putting well requires a careful blend of good mechanics and a deft touch.

**FIG 4-62** Practice Flop, shots from deep grass.

**FIG 4-63** Chipping around the box.

143

**GOOD PUTTING MECHANICS** To improve the mechanics of your putting, start by paying careful attention to the alignment of your body over the putt and the swing path of the club head into the back of the ball:

1. Set up with your feet, hips, and shoulders parallel to the Plane Line. The shoulders are the most likely to be "open" because you can't see your shoulders when setting up.

2. Using the Swing Light, train yourself to swing the head of the putter relative to the Plane Line.

It is the path of the putter head into the back of the ball that dictates the direction the ball will roll.

You can train yourself to swing the putter head squarely into the back of the ball. The Swing Light Trainer will let you know immediately if you are out of position before you hit the ball. Too often you are looking at a spot out in front of the ball to get the

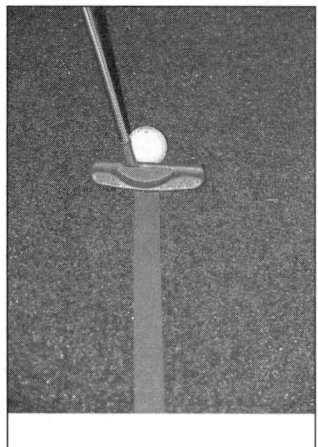

**FIG 4-64**

**FIG 4-65**

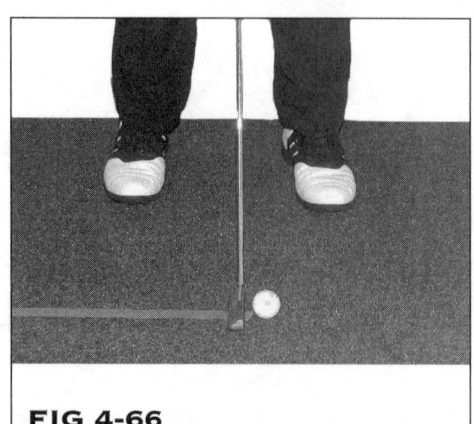

**FIG 4-66**

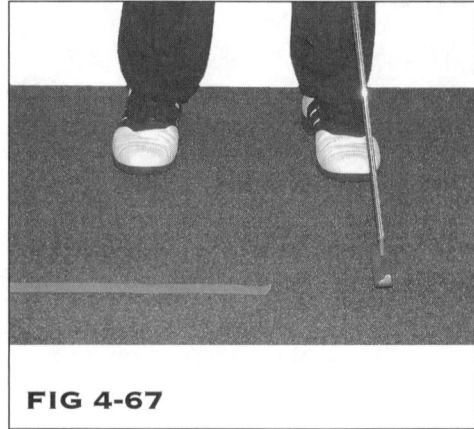

**FIG 4-67**

putt rolling on-line. By then it's too late. Focus your training on the path of the putter head into the back of the ball.

**DEVELOP A DEFT TOUCH** It is difficult to improve your "feel" for putting. When I talk about feeling in putting, I am talking about how hard to hit any given putt. To train for feel, try looking at the hole instead of the ball when you practice putting. This requires trust. Just remember, when you shoot a basket, play catch with a baseball, or throw a football, you're looking at your target (Figure 4-68). Why not in putting? Perhaps golfers miss putts because they can't remember the location of the hole.

**TRAJECTORY, TEMPO, AND TIMING** Would you agree it's easier to hit your 7-iron 150 yards than it is to hit your 3-iron 200 yards? Do you realize the difference in ball flight is only 50 yards? Have you ever tried to hit a golf ball exactly 50 yards? It's hard to do because it's less than a full sand wedge.

Try this drill on the golf course, not on the driving range: place a ball at the 200-yard marker in the middle of the fairway and hit three shots to the 150-yard marker. You will discover it takes very little effort to advance a golf ball 50 yards.

The difference between your PW and your 7-iron is about 50 yards of ball flight. The PW travels about 100 yards, the 7-iron travels about 150 yards.

The difference between your 3-iron and your driver is about 50 yards of ball flight. Your 3-iron travels 200 yards, your driver travels 250 yards.

**FIG 4-68**

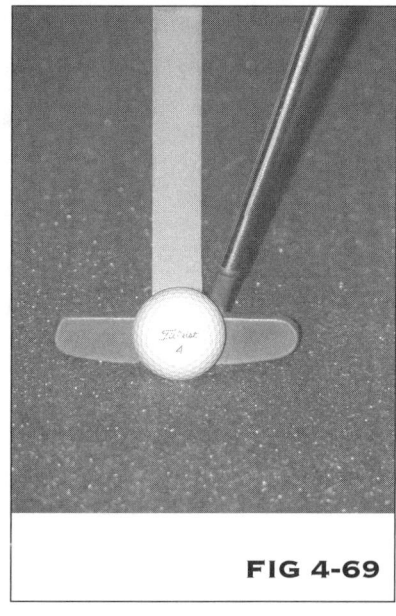

**FIG 4-69**

You can discover, with trust and patience, that how far you hit the golf ball has little to do with physical strength. It has to do with the loft angle of the club in your hand.

> P-wedge: 48 degrees of loft = 100 yards
> 7-iron: 36 degrees of loft = 150 yards
> 3-iron: 21 degrees of loft = 200 yards
> Driver: 10.5 degrees of loft = 250 yards

**LOFT ANGLES** Shift your awareness then to the height of your shots off the club face. Use the same "effort" with your 7-iron that you use with your 3-iron. Remember, the difference is only 50 yards . . . a very short distance.

Imagine this: the effort you use to advance the golf ball 100 yards with your PW is the same effort required to advance the ball 250 yards with your driver!

If you feel the need to swing harder or faster with the longer clubs in your bag, limit your effort to the different length and weight of the clubs in your bag. The driver is about 10 inches longer than the PW; therefore, it is a bit heavier.

Hitting the ball farther, then, is a matter of trajectory of the ball flight based on the loft angle of the club. Let your tempo remain constant. Let your effort notch up a bit as the clubs get longer and slightly heavier. Let the loft angle of the club face do the work (Figure 4-70).

## WARNING!

When all is said and done, you might discover you enjoy training for golf more than playing golf for score. When you train, four different stats are on the line for every hole.

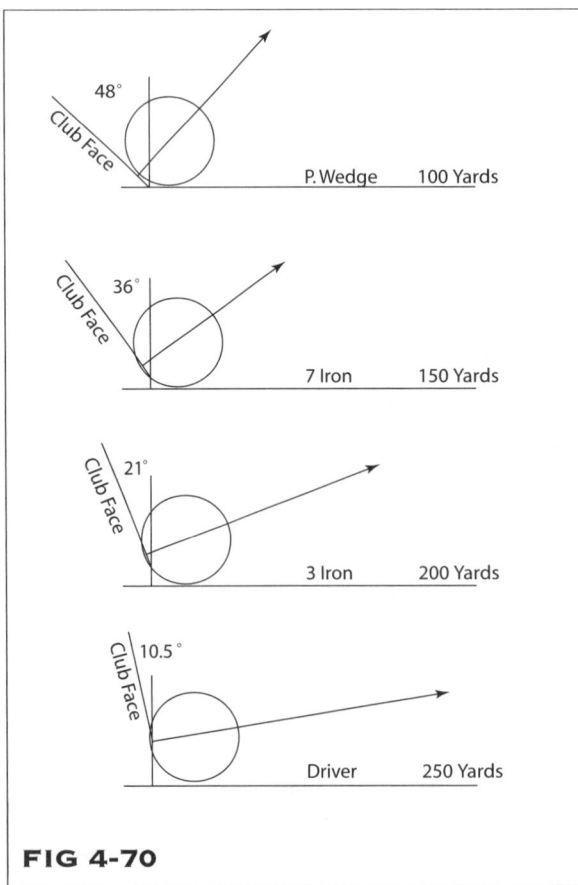

FIG 4-70

Here is an example of how to train for golf on a typical 380-yard par 4:

### If you shoot in the 100s:

Tee ball to the 225-yard marker requires a 155-yard shot.

Second shot of 125 yards to the 100-yard marker.

Third shot into the green.

### If you shoot in the 90s:

Tee ball to the 200-yard marker requires a 180-yard shot.

Second shot of 100 yards to the 100-yard marker.

Third shot into the green.

### If you shoot in the 80s:

Tee ball to the 175-yard marker requires a 205-yard shot.

Second shot of 75 yards to the 100-yard marker.

Third shot into the green.

### If you shoot in the 70s:

Tee ball to the 125-yard marker requires a 255-yard shot.

Second shot into a specific quadrant on the green.

At first it might seem boring or unnecessary to hit these short shots. However, once you take on the challenge of actually getting within 20 feet of your target on these shots it becomes fun. In time, you will learn these shots of 100 yards or less are an essential part of your training for lower scores.

Remember, the ultimate goal from 100 yards out is an "up and down." Hit it close and make the putt. If you miss the green, hit it close and make the putt.

**ROLL IT IN THE CUP** When putting, once you determine the best line to the hole, train yourself to look at the Plane Line behind the ball. Take your practice swings moving the putter head into the back of the ball along the Plane Line. Then, train yourself to trust your mechanics by looking at the cup when you putt. If you train this way, putting "for real" will seem easy by comparison.

There are infinite combinations of training strategies inside the model I have presented. The key is your commitment to training. Should you decide to train for stats the next several years instead of playing golf for score you will become a better player.

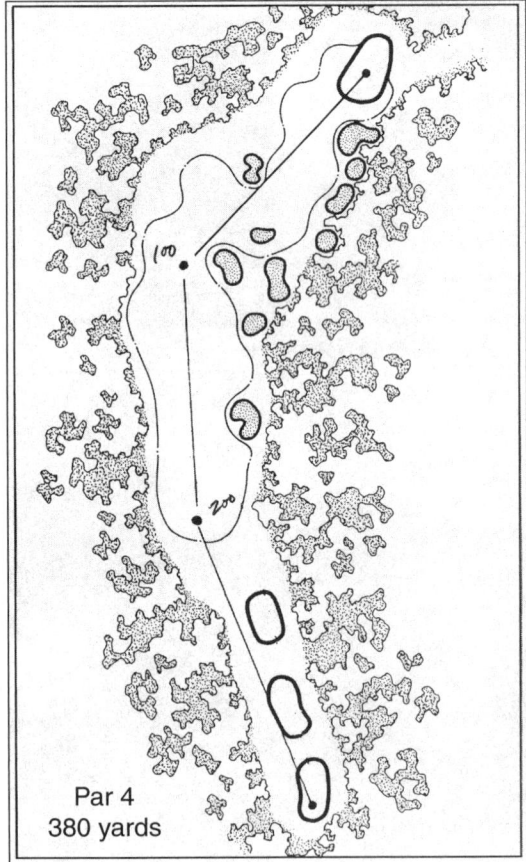

Par 4
380 yards

**FIG 4-71** Training sequence for players who shoot in the 90's.

*Source:* © 1993, Robert Trent Jones, Jr., *Golf By Design.*

# Chapter 5

## FEAR AND FREEDOM

### The Golfer's Guide to Mental Health

You and the Ball, You and the Target

Satisfaction

Freedom

Acceptance

Polarity

How the Mind Works

The Challenge of Golf

Unnecessary Interference

Practical Applications (Drills)

# YOU AND THE BALL,
# YOU AND THE TARGET

Have you had the experience of driving your car, and then, suddenly, you lose visibility because of fog, heavy rain, dust, or smoke? All at once, you feel panic. You immediately slow down. You become nervous and tentative. Your focus has shifted from the road out ahead to the asphalt just beyond the hood of your car. Now, each and every tiny movement of the steering wheel feels critical to keep the car on the road.

Have you had the experience of playing a great round of golf, and then, suddenly, you're looking down the narrow fairway of a long par 4 with out-of-bounds stakes on the left and a lake on the right? All at once, you feel panic. You immediately try and slow everything down. You become nervous and tentative. Your focus has shifted from the landing area that is 250 yards down the fairway in front of you to the many different movements your body needs to make to produce a good swing. Every thought, you know, is critical to remember and to rehearse.

You can drive your car in a state of physical freedom as long as you can see the changing conditions of the road from a distance. When your focus shifts from the road out ahead to the asphalt just beyond the hood of your car you are driving in a state of conscious defensiveness to avoid an accident or injury.

You can swing a golf club in a state of physical freedom as long as you can hold the image of the target in your mind's eye when you swing the club. When your focus shifts from the target to the many swing thoughts in your head you are playing golf in a state of conscious defensiveness to avoid poor performance.

## SATISFACTION

I have yet to meet a golfer who is satisfied with his or her game. Every golfer wants to play better. The 25 handicapper wants to break 90. The 12 handicapper wants to break 80. The tour pros need to lower their scoring average to avoid losing their playing privileges for next season. Golf's famous elite are trying to play better to qualify for a spot on the Ryder Cup Team or the Tour Championship at the end of the year.

*Are you satisfied with your level of play?*

The point is, no matter what your level of play, you want to play better. If you've been at the game for several years, you're already deeply invested. You know you're doing everything you can to improve. You practice. You take lessons. You make frequent equipment changes with clubs, balls, gloves, and shoes. You read books and watch videos of golf instruction. You've even got the Golf Channel on most everyday.

In short, you're frustrated. It seems for every great shot you experience there are 10 forgettable shots. Let's face it, your game is not the game you hoped it would be. Most people would have given up long ago, but not you. Why do you still practice? Why do you still play? Maybe it's because, deep down inside, you know you're not the problem. You have reasonable athletic ability and decent hand–eye coordination. You have objective proof that there are men and women who play the game at or very near par every time they tee it up. There are men and women found all over the world that play the game so well, so effortlessly. They are flesh and blood mortals, just like you and me, yet they must know "something" about the game of golf that you don't know. They must have found "something" that you haven't found. If you could find that "something," you know you could play!

## FEAR

When I play golf, I fear looking foolish. I'm certain that statement would keep me busy with a psychotherapist for many years, but maybe my fears are your fears. I fear driving the ball into the grove of trees on the right. I fear hitting my approach shot into the deep-faced bunker on the short side of the green. I fear skulling my bunker shot into the lake. I fear missing the 3-foot putt to save par. And I'm a pretty good player! I compete professionally. My scoring average is 72. My career low round is 62. It doesn't seem to matter, I'm still a wreck when I play and it only gets worse when I compete.

> Fear is nothing to be afraid of.

Away from golf, I can stand at the free-throw line and release the basketball towards the hoop. I might not make 'em all, but I know I'll at least hit the rim or the backboard. I won't throw air-balls and miss everything by 5 to 10 feet. I can stand 90 feet away and play catch with my son. When I release the baseball, I know he's going to be able to catch it. I know I won't throw the ball over his head or so far left or right that he can't reach it. I can stand on a tennis court and sustain a back-and-forth rally of the ball over the net for a reasonable period of time. When I release the tennis racket into the back of the ball, I know I'm going to hit the ball somewhere toward my target inside the green paint on the other side of the net.

In golf, however, I can swing my driver and miss a 30-yard-wide fairway by 30 yards! I can swing my 6-iron and miss the entire green by 30 feet! I can hit a 10-foot putt and miss the hole by 3 feet! Why is golf so difficult?

The answer is based, in part, on the principle of leverage. When you shoot a basketball or throw a baseball, your starting with the ball in your hands. The physical length of your arms, from your shoulders to your hands, is the lever. Your "levers" are attached to your body and have millions of nerve endings signaling awareness of their motion and movement. By comparison, when you hold a tennis racket or swing a baseball bat, you instantly lengthen your lever by about 24 to 36 inches. As the levers get longer, they are more difficult to control.

Golf clubs are the longest levers in ball-oriented sports. From your shoulders to the end of the driver can be as long as 5 feet! Nevertheless, there are about 500,000 men and women around the world who can wield these levers as if they were a natural extension of their arms and hands. Sure, they miss short and long, left and right, but not near as far off-line and not near as often as we do. So, it can be done. Make no mistake about it, it can be done.

## FREEDOM

I have dreams of playing golf with freedom. Freedom from fear of failure. Freedom from embarrassment in front of playing partners or the gallery. The dreams are always the same. I see the target. I select the appropriate club. I swing and release the club head effortlessly into the back of the ball. The ball flies exactly as I imagined it would to the target. When I look down a fairway, all I think about is the landing area I will hit. When I look at the green, all I see is the area of placement for the ball to land to give me the best chance for birdie. When I putt, all I am aware of is the best line and pace of the ball to roll into the bottom of the cup.

**FIG 5-1** Without conscious thought about her moving body and the position of her pole, the trained athlete always holds the target (the cross-bar) in her mind's eye.
*Source:* © Martin Gushen/Alamy Images.

By comparison, whenever I shoot a basketball, I see only the net. When I throw a baseball, I see only the glove in the left hand of my son. When I hit a tennis ball, I am aware only of the placement of the shot over the net inside the green paint. I am completely target oriented. It would never occur to me to think otherwise.

Yet, when I swing a golf club, the target is usually the last thing on my mind. Sure, I noticed the target, but by the time I swing, many, many thoughts have superceded the target. And each and every nontarget thought tightens the muscles of my body, one by one, until I am anything but free to release the club head into the back of the ball.

## MAKE YOUR CHOICE

Why not make a conscious choice to play golf with the same freedom you play other ball-and-lever-oriented sports? You have proof there are many men and women who are doing just that. Why not decide to be one of them?

# ACCEPTANCE

## THE PAST

Golf is frustrating. However, have you ever considered how helpful it has been to be frustrated? You say, "What?! There is nothing helpful about frustration." To which I suggest, you will never appreciate the freedom of a confident golf swing unless you know the contrast of a tight and fearful golf swing. Contrast has been a necessary experience for you. You would never appreciate the great golf shot unless you've first experienced golf shots that were horrible and embarrassing. The contrast of your past experience can be the greatest gift in your life because knowing what you don't want is just as important as knowing what you do want.

*Decide to appreciate contrast.*

## THE PRESENT

Where are you now? What is the current state of your game? How many fairways do you hit in regulation? How many greens do you hit in regulation? How often do you get up and down when you miss the green in regulation? How many putts do you average per round? Without answers to these questions you might as well be trying to get an airplane into the sky without wings on the fuselage. Your stats are critical indicators of the health of your game. When you commit to keeping and objectively interpreting your training stats, you are holding to your commitment of improving your golf game. Stats can guide you around the golf course with better club selection. Stats can guide you in your practice sessions away from the golf course.

If you are unfamiliar with the technology of stats, everything you need to know has been carefully explained in Chapter 4 of this book under the heading "The 9-Hole Playing Lesson."

> What percentage of fairways do you hit in regulation?
>
> What percentage of greens do you hit in regulation?
>
> How often can you get up and down when you miss a green?
>
> How many putts do you average over 18 holes?

## THE FUTURE

Where do you want to be one year from today? Set a goal. Make sure your goal is realistic based on the overall physical condition of your body and the amount of time you have to play and to practice each week. Make sure your goal is easily and objectively measurable. (Stats are perfect for this.) Make sure your goal includes the component of time. If you've made a one-year goal, set interim goals along the way to evaluate your progress. Your goals can be physical, such as a better scoring average, or your goals can be mental, such as hitting every shot with freedom.

An example of a physical goal: I will lower my handicap from 18 to 9 by one year from today. The goal is realistic if you commit to everything you learned when reading this book. The goal is objectively measurable using stats. The goal includes the element of time, "within one year."

An example of a mental goal: shot by shot, I will swing the club when I feel fearless. The goal is realistic. (You've got to keep reading to understand how, but you can do it.) The goal is objectively measurable. Of the 82 shots in your round, what percentage where made while feeling fearless? The goal has the component of time. It's right now! It's every shot.

You must set your goal to reach your goal.

So, it's up to you. If you want to swing the club with freedom and give yourself the chance to play your best golf, then keep reading.

# POLARITY

If you're going to have any chance of playing better golf, you're going to have to learn to chase those fearful thoughts from your head. Before you can begin, you've got to understand the anatomy of thought. You need to understand polarity.

By definition, *polarity* means, "possession or manifestation of two opposing attributes." The earth, for example, has a North Pole and a South Pole. Each pole has specific and opposing electromagnetic attributes that are essential to maintain balance of the ecosystems and therefore the survival of the planet. The human body is composed of positive and negative electrons. Electrons that have opposing attributes attract each other, bond together into different compounds, and sustain the form, the function, and therefore the life of your physical body. In the nonphysical world, humans and animals alike sense, feel, or respond to only two opposing emotions: love or fear. Each emotion is characterized by opposing attributes. Creatures are drawn to love and run from fear.

In every circumstance of your life, polarity is a constant. The planet you live on, the body you live in, and the emotions you feel are all tied together by polarity. And—are you ready for this? Your golf swing is polar. Before you start to swing, you are either swinging with fear or you are swinging with freedom. You ask, "Who would choose to swing with fear?" Unfortunately, about 99.5 percent of all the golfers of the world swing the club in a state of fear.

When you shoot a basketball at the hoop, you shoot with freedom. When you throw a baseball, you release the ball from your hand to the other player with freedom. When you swing a tennis racket, you hit the ball across the net with freedom. When your thoughts are target oriented, you are free. All you really see is the hoop, the glove, or the green paint on the other side of the net. You are not bound up with mechanical thoughts of positional alignment of your hips, shoulders, arms, and wrists. Far from it! Mostly, you're in conversation with your recreational partners. There are other people talking and running about. There are all kinds of different noises and potential distractions that you do not see and do not hear. In golf, we call that state of mind being "in the zone." In every other sport, we call it the natural state of being.

Certainly, competitive and professional basketball, baseball, and tennis have pressure that athletes must learn to deal with. Have you ever seen the fans sitting behind the basketball backboard waving banners while shouting and screaming "Miss!" as the player

> In every circumstance of your life, polarity is a constant.

PART 2 – FOR THE PLAYER

attempts to make a freethrow? Fortunately, for 99.5 percent of the nonprofessionals who play golf, that kind of pressure does not exist. Except in your mind.

## THE POSITIVE

Hit the ball in the fairway. Hit the ball onto the green. Putt the ball into the hole. These are examples of strong and decisive thoughts. They are positive. Remember, your thoughts are pure energy. You can choose to hold positive thoughts.

*You'll have to learn to think better to play better.*

## THE NEGATIVE

Don't hit into the trees on the right. Don't hit into the pond in front of the green. If I miss this putt, it will cost me $5. These are examples of fearful thoughts. They are negative. Remember, your thoughts are pure energy. You can choose to hold negative thoughts.

*Don't say don't.*

## THE WHOLE

It is impossible to avoid thinking negative thoughts. However, it is possible to not let negative thoughts take hold and disrupt your freedom. Never forget, you are polar. You are both negative and positive all at once. What you choose to hold in your mind at any given time is a choice only you can make.

Imagine standing in the fairway contemplating a shot that has to carry 145 yards to the center of the green that is guarded by a small pond. At some point, you will think and you will feel the fear of the ball landing in that pond. You will consider putting an older ball into play, you know . . . just in case. You think about the effect that a bad swing here would have on your score and the bet you have with your playing partner. These thoughts are like bees that begin to swarm in your head. First, there are one or two, but by the time you swing the club, a whole colony of bees is looking for the queen that landed on your collar.

Or you can offer no resistance to the negative thoughts. You can remain calm inside. The negative thoughts will pass, if you let them. You can remember you are polar. Eventually, you can imagine the queen bee has flown away and all those pesky bees have followed her to your playing partner. You can choose to hold a positive thought when you swing: "Hit the green." When you can hold a positive, target-oriented thought when you swing, you are playing with freedom.

*To play better, choose better thoughts.*

# How the Mind Works

This sounds too simple, I know. If it were so easy to side-step fear and swing with freedom, everyone would do it, right? Not exactly. In my experience, the golfers I have worked with are not in a state of consciousness about the state of their mind when they play golf. Only when asked, and it usually takes a while to find the answers to my questions, do players realize they might have a choice about what they hold their attention on when they swing the club.

*What are you thinking about when you swing the club?*

## Channels You Like

The thoughts in your mind are like different channels on your TV or stations on your radio. You switch from channel to channel until you find something you like to watch. You switch from station to station on your radio until you find something you like to hear. Television signals and radio waves are electromagnetic energy. You choose to tune them in, and you can choose to tune them out. The thoughts you hold in your mind are also electromagnetic energy. You can choose to hold a thought. You can choose to let a thought go.

*It's your mind. It's your choice.*

## Channels You Don't Like

When swinging the club, most players hold thoughts they don't like. It's not an intentional process. It's an unconscious reflex based on past failure. It began the first time you surveyed the multitude of hazards on the golf course. There were trees *not* to hit into, there was water *not* to hit into, and there were sand traps *not* to hit into, and on and on. From day one, and usually with help from your playing partners, you have unconsciously chosen to hold thoughts of fear of hitting into hazards when swinging the club. With best intentions, before you swing, you advised, "Watch out for that lake on the right." Or, "Whatever you do, don't hit into that trap by the green. The sand is really soft and really deep. You'd be lucky to ever get out." You cannot swing with freedom when your thoughts are tuned into K-TREE or K-LAKE or K-SAND.

*Are you on the right channel?*

## Changing Channels

Because your natural state is one of polarity, the channels and the stations in your mind are loaded with programs featuring your life's experiences of fear and freedom. Your mind cannot avoid channel surfing. You see stuff you like and stuff you don't like . . . whether you like it or not. You do have the power, however, to change the channel before you swing the club. Becoming aware that you are on a channel you don't like is a huge first step. In time, as you are preparing for your up-coming shot, you will become amused at the fearful thoughts that present themselves for consideration. Like annoying advertise-ments, they beg for your attention. The key is to say no to fear. The key is to choose the better thought, the thought that allows you to swing the club with freedom.

*It's easier than you think.*

## THE CHALLENGE OF GOLF

Playing a really good round of golf demands mastery of an ability to trust yourself, the willpower to concentrate, and the commit-ment to practice patience. When you decide to devote your atten-tion to these issues, you decide to take on the greatest challenges in all of sport, in all of life.

*Golf is more than golf.*

### TRUST

*Trust* is defined as total confidence. You cannot swing the club with freedom if you don't feel confident. Golfers struggle with confi-dence over almost every shot. And with good reason. Golfers use the longest levers in sport. In addition, there are 14 different length lever arms (club shafts) to choose from, depending on the shot at hand. On average, the distance from the shoulder to the club head at the end of the driver is more than 5 feet! Regardless of your pre-golf athletic abilities, swinging these different length levers at a tiny golf ball that sits so very far away takes some getting used to. To make matters worse, for more than three-quarters of the swing, you cannot see the lever when the club head is wrapping around and behind your body. In short, golfers must learn to play with trust.

*Do you swing the club with confidence?*

### CONCENTRATION

*Concentration* is defined as undivided attention. Golf demands you hold your attention, undivided, on the target when you swing the club. It takes less than 2 seconds to swing the club, yet the average golfer processes five different thoughts each time he or she swings the club. Thought 1: your swing key for the shot at hand, for example, "Keep your left arm straight or finish the backswing." Thought 2: this one occurs at the top of the backswing, for example, "OK, now hit the ball!" Thought 3: this one occurs during the downswing when you say to yourself, "You know, something doesn't feel right." Thought 4: this one occurs at the moment of impact when your thoughts immediately process the feeling of impact, for example, "That shot was pure, or thin, or fat." Thought 5: this one occurs before you have finished your follow-through as you evaluate the flight path of the ball, for example, "That damn ball is slicing hard right. Fore!"

In addition, a round of golf takes about 5 hours to play. Can you concentrate on anything for 5 hours? Can you name another mainstream sport that asks so much over such a long period of time from those who play? Factor in the issues of dehydration (from too much soda or alcohol), fluctuating blood sugar levels (from junk food), and environmental allergens (pesticides and fertilizers on the grass) and it's no wonder that the worldwide average score for 18 holes of golf is a staggering 28 strokes over par. If the average golfer played in a PGA tour event, after 4 days, he or she would finish well over 100 shots behind the winner of the event! Even if you play to a 10 handicap, you'd get beaten by more than 40 shots after four days.

## PATIENCE

*Patience* is defined as the capacity to endure affliction calmly. To be patient is to quietly suffer. Golf demands patience, not in the expected form of tolerating slow play, but in the unexpected challenge of the training necessary to play the game. In general, we just want to play golf. We do not have the patience to train for golf. Make note: taking golf lessons is not the same as training to play golf.

To play better golf, to actually improve, you have to be willing to stop keeping score and start keeping stats. (See Chapter 4, "Training for Golf.") When you train for golf, there are four different stats to play for on every hole: fairways in regulation, greens in regulation, your percentage of up and downs whenever you miss the green, and putts per green. Commit to improve your stats and you will improve you score, dramatically!

**FIG 5-2** Pure concentration on the target . . . the glove of the catcher.
*Source:* Bryan Eastham/ ShutterShock, Inc.

Keeping stats can test your patience. If you are committed, for example, to putting your tee ball in the fairway, you must be prepared to leave your driver in the bag. Of course, if you can hit your driver into the fairway 70 percent of the time (that's 11 of 14 fairways per round of golf), then swing the driver. When I committed to improving my stats, I discovered the only club I could hit into the fairway 70 percent of the time was the 5-iron. It took the next 18 months for me to progress to hitting the driver off the tee. Along the way, it took every ounce of patience I could muster not to pull the driver out of the bag and swing away on every par 4 and every par 5.

Long ago, it was written: patience is a virtue. In your quest to play better golf, patience is a requirement.

> To be patient is to quietly suffer.

# UNNECESSARY INTERFERENCE

To play golf with trust, concentration, and patience you must have the electromagnetic field of your mind in a state of "clear." A clear mind is a mind capable of processing and evaluating the literally thousands of bits of information you need to consider during 18 holes of golf. Unfortunately, there are potential barriers to becoming and/or maintaining a state of mental clarity.

> Good golf requires good decisions.

### EMOTIONAL

Stress can ruin your ability to play golf with trust, concentration, and patience. Although it is beyond the scope of this text to treat the underlying cause of the stress you have in your life, I can offer a solution to the symptoms of stress: homeopathic remedies are an excellent nutritional supplement to combat and neutralize the effect of stress in your body. Homeopathics are not the cure for stress, but homeopathics can get you through a round of golf.

ER911 and Rescue Remedy are two examples of homeopathic remedies for stress that can be purchased without prescription at any health food store. Just a few drops under the tongue about 30 minutes before your round begins and you'll feel your body relax and be ready to go on the first tee. During your round, the homeopathics are helpful when you get angry after a bad shot and they are helpful when you get goofy after a good shot. Either way, their calming effect will benefit your ability to play your best golf.

Information about homeopathic remedies for golf can be found by visiting www.GolfInjurySeminars.com.

## ENVIRONMENTAL

Unfortunately, golf courses are loaded with pesticides and fertilizers that are harmful to most everyone who plays golf. In addition, golf courses are loaded with different types of grass, foliage, and fauna to which you might be allergic. Whenever your body comes in contact with allergens, the muscles of your body tighten, by unconscious reflex, and prevent you from swinging the club with freedom.

Homeopathics can offer a simple solution. Again, homeopathic remedies will not cure your allergies, but homeopathics can combat and neutralize the effect of an allergic reaction in your body.

Allergy remedies can be purchased without prescription at any health food store. Keep the bottle in your bag. Place several drops under your tongue about every third hole. There are no harmful effects from homeopathic remedies. They cannot hurt; they can only help.

## NUTRITION

Hydration is essential for brain function. Without adequate water in your body, the electromagnetic signals that pass through your brain slow down. Your thinking can get foggy.

To adequately hydrate your body, be sure and drink half the number of your body weight in ounces of water each day. Soda, tea, coffee, and alcohol *do not count!* In fact, they dehydrate your body further.

A word about junk food: the blood sugar level in your body is dramatically influenced by carbohydrates. Initially, after eating foods rich in carbohydrates your blood sugar level spikes and you feel energized. Unfortunately, within 45 to 60 minutes, your blood sugar level will crash. You will feel tired, like you need to take a nap. Again, your mind will become foggy.

To avoid the blood sugar rollercoaster, try to eat more protein-oriented foods and drink plenty of good clean water. Protein provides

**FIG 5-3** Stress is elation. Stress is anguish. Your central nervous system cannot differentiate between the two.

*Source:* © Eric Miller/ Reuters/Landov

Better nutrition is needed for better golf.

fuel for your body that burns at an even rate, supplying your brain with nutrition that could be considered more stable.

# PRACTICAL APPLICATIONS (DRILLS)

Swinging the club with freedom begins with an awareness that your mind is filled with thoughts of fear. Once you've become aware, you

The first step toward change is awareness.

must learn to accept that you are, by your very nature, polar. That fear and freedom coexist. That you cannot experience freedom without the experience of fear. Once you embrace this truth, you are ready to make a choice: Am I going to swing the club with fear, or am I going to swing the club with freedom?

When you choose to hold the target in your mind, you'll find it easier to swing with freedom because the targets, and your troubles, are far away. This is like looking at a beehive from a

Every swing is another opportunity to develop your ability to concentrate on the target.

safe distance. You have no fear. You are at peace. When your mind is holding thoughts of fear, your troubles, like a swarm of angry bees, are upon you. The very thought of a venomous sting binds the muscles of your body to restrict the release of the club head into the back of the ball.

## ON THE DRIVING RANGE

Before you hit your first ball, decide you are going to use this time to train your mind and your body to play better golf. As you swing the club around your body to loosen up, pick out a target on the range. Maybe it's a red flag. As you take your practice swings, try to hold the image of the red flag in your mind. If any other thought breaks your concentration, understand your mind has changed to another channel. Go back to the "target channel" and stay there as long as you can, swing after swing.

Next, put a ball in play. Again, decide on your target, lock it down in your mind, and swing away. If another thought breaks your concentration, understand your mind has again changed the channel. Be patient with yourself. Consciously switch back to the "target channel." With time, these drills become easier. Remember, your mind, like any other muscle in your body, needs exercise to function at optimal health.

## ON THE GOLF COURSE

If you are committed to improving your stats, you will have no trouble picking out specific targets when you are out on the course. However, the challenge to stay locked on target becomes magnified away from the static conditions of the driving range.

On the driving range you have the same flat lie of ball on the ground—exactly the opposite of when you are on the course. You repeatedly hit toward the same, well-defined targets over and over—exactly the opposite of when you are on the course. Mix in the variables of weather conditions and the distractions presented by your playing partners, and it's easy to break your concentration when you are on the course.

Your preshot routine, described in Chapter 4, "Training for Golf," is essential to bring you back to focus when faced with the distractions on the golf course. Don't get discouraged. Make it a game within a game. Let your mind and body work together to produce golf swings that are free from fear.

> Your preshot routine should focus your concentration on the target.

## IN EVERYDAY LIFE

When you walk, you have a destination (target). Maybe it's walking through a parking lot to your car. You see the car in the distance, and your body moves forward step by step until you reach your destination (target). You don't have to think about moving forward in a specific direction, you just do it. Now imagine looking down as you walked trying to remember where you parked the car, putting your focus on each and every step along the way. By changing your focus from your destination (target) to your body, you can make the almost effortless task of walking to your car a complicated mess as you try to place purposefully one foot in front of the other. In an instant, you can again change your focus away from your body and onto the target and regain your freedom of movement.

As you drive your car, notice how far ahead you set your eyes down the road in front of you. Now change your focus to the road just over the hood of your car. Notice the additional steering and intensity you feel in your body as the road appears to come upon you so quickly. In an instant, you can change your focus to the road far ahead and immediately relax and drive with a feeling of freedom.

When you play golf, keep your targets far away and off in the distance whenever possible. The moment you feel fear, the moment you begin to feel tight, notice you have changed your focus away

Practice, every chance you get, to concentrate.

from the target and onto your body or onto the ball or both. Before you swing the club, make a conscious choice to change the channel in your mind to the target in the distance and swing with freedom.

Practice putting looking at target. Once you're on the green, the target gets closer and closer to your body with every stroke. The physical distance of your body from the target makes it easier to feel freedom when your swing the club. Therefore, as you get physically closer to your target, the harder it becomes to feel freedom. Primarily, this is because you have a belief that goes something like this, "I'm so close, and it would be embarrassing to miss this short putt." You might find some relief in knowing that tour professionals only make 50 percent of attempted putts from 5 to 10 feet away from the cup; 50-50. That's means they miss, they make, they miss, they make, they miss, they make. . . .

You can just about guarantee you'll miss when you let your mind leave the target when you swing the club. If you let your thoughts switch to awareness of your body, your putter, the ball, or the consequences of whether you make or miss, you are doomed.

You will have a better chance of making putts if you can switch your concentration to something inside the hole itself. Any scuff mark or lettering you can see inside the cup will do. Lock onto that target and swing with freedom. If you do, when all is said and done, you will make more than you miss.

# AFTERWORD

Implementing change is often difficult. Learning new information is one thing. Putting that information into practice is quite another. Sooner or later, someone has to be your *first* golf injury patient.

Let's take a look at how tour pros implement a change into their golf swing and see how this process might apply to you as you begin to treat injured golfers with the new information you have learned in this textbook.

First, there is the element of time. Change will take time. The intention to make a change is step one. However, there will be many tiny steps along the way before you are comfortable with change.

The pros begin the process of swing changes on the driving range, away from the pressure of tournament play. You should begin by selecting a few family members or close friends as your first patients and practice getting the feel and flow of the golf-specific examination procedures. You need time to practice your technique and your explanations to patients about how the various correction protocols relate to their golf swing.

Once a tour pro feels they can implement the swing change with some consistency on the driving range, they will take the changes to the golf course during practice rounds. For you, now is the time to begin treating patients from inside your present practice. Select a few non-family, non-friend patients who already know you, and who already trust you. I recommend that you continue to treat patients from inside your present practice until you begin to get referrals.

Now, you are treating patient referrals coming into your office from outside your practice. In time, you will develop your skills and confidence in the diagnosis and treatment of many different golf patients with many different kinds of conditions. After a tour pro has had the time to make cahnges in their swing during practice rounds, he/she will decide to debut the new swing under the pressure of tournament conditions. For you, a similar debut means you are now ready to approach the PGA instructors in your community and present your expertise to your community as a golf injury specialist. Once the PGA instructors recognize the value of your service, referrals from them are guaranteed follow.

# PRODUCTS AND SERVICES

Information about Seminars

Find a Golf Injury Doctor

# FOR INFORMATION ABOUT:

Golf Injury Training Seminars
Swing Light Trainers
Play On Pain Relief Lotion

please visit my Web site at:

www.GolfInjurySeminars.com
Phone: (805) 772-8298

# FIND A GOLF INJURY DOCTOR

**GOLFINJURYDOCTORS.COM:** GolfInjuryDoctors.com is an online directory of health care providers who have completed post-graduate training in the diagnosis and treatment of common golf injuries.

**GOLF INJURY CERTIFIED:** Many of the doctors have taken additional advanced training in the management of "well" golfers. These doctors are trained to help men, women, and children who are interested in golf-specific fitness and conditioning for improved performance on the golf course.

Find a golf injury doctor online or by telephone at (805) 772-8298 for a referral.

# REFERENCES

Chek, Paul. 1999. *The Golf Biomechanics Manual.* Vista, CA: Chek Institute.

Egoscue, Pete. 1992. *The Egoscue Method of Health Through Motion.* San Diego, CA: HarperCollins.

Kapandji, I.A. 1974. *The Physiology of Joints.* New York: Churchill Livingstone.

# ABOUT THE AUTHOR

Dr. Jeff Blanchard has been a chiropractor since 1980. He has produced original equipment design and modifications to the Percussion Adjusting Head for upper cervical chiropractors who adjust the spine by machine. He developed Pelvic Hip Calipers for the standardization and objective measurement of patient leg-length discrepancies. He is the founder and developer of the Whiplash and Spinal Trauma Center in San Diego, which has become the model for hundreds of similar facilities throughout North America and Seoul, South Korea.

In 1987, Dr. Blanchard began his intensive training in the field of golf. Having dismissed the traditional model of golf instruction, he began to use his years of training in applied geometry and human biomechanics to create a new model of golf instruction.

After 12 years of intensive research and development, he sold his practice and devoted himself entirely to playing professional golf and to teaching about his discoveries, techniques, and methodologies. Since 2000, he has attracted more than a thousand students each year to his seminars.

Dr. Blanchard lives in San Luis Obispo, California, with his wife Chantal. He has three children: Erin, Alexandria, and Jean-Marie.

Page numbers followed by *f* indicate figures.